CW01494813

The Imp
Roman Army

30BC–AD284

History, Organisation and Equipment

GABRIELE ESPOSITO

KEY
Books

HISTORIC ARMIES SERIES, VOLUME 2

Contents page image: Roman legionaries advancing on the battlefield in line. (Legio XI C.P.F. Hispaniensis)

The author

Gabriele Esposito is a military historian, author and researcher in the military history sector. He specialises in uniformology. His interests and expertise range from ancient civilisations to modern post-colonial conflicts. During recent years, he has conducted several lines of research on the military history of non-European countries. His books and essays are published on a regular basis by international publishers. Gabriele is also the author of numerous military history articles appearing in specialist magazines.

Acknowledgements

This book is dedicated to my magnificent parents, Maria Rosaria and Benedetto. A very special thank you to Jonathan Jackson and Brianne Bellio of Key Publishing for believing in my editorial projects. Their love for history and passion for publishing were fundamental for the birth of this book.

A very special mention goes to the brilliant re-enactment groups that collaborated with me, allowing their photos to be used in this book: without their incredible research, this publication would not have been the same. Specifically, I want to express my deep gratitude to the following living history associations: Ala I Batavorum from the Netherlands (especially Jurjen Draaisma), Legio II Traiana Fortis – Cohors I Barcinonum, Barcino Oriens from Catalonia (especially Marc Seriol), Legio XI C.P.F. Hispaniensis from Spain (especially Iago Rodriguez Diaz), Legio VI Victrix Cohors II Cimbria from Scandinavia (especially Thomas Noe), Legio XIII Gemina from Austria and Legio XIIII Gemina from the USA (especially David Burns).

Published by Key Books
An imprint of Key Publishing Ltd
PO Box 100
Stamford
Lincs PE9 1XQ

www.keypublishing.com

The right of Gabriele Esposito to be identified as the author of this book has been asserted in accordance with the Copyright, Designs and Patents Act 1988 Sections 77 and 78.

Copyright © Gabriele Esposito, 2023

ISBN 978 1 80282 593 0

All rights reserved. Reproduction in whole or in part in any form whatsoever or by any means is strictly prohibited without the prior permission of the Publisher.

Typeset by SJmagic DESIGN SERVICES, India.

Contents

Introduction

I aim to present a detailed overview of the Roman Army's history, organisation and equipment during the period 30BC–AD284. My analysis begins with the year 30BC, which saw the ascendancy of Octavianus Augustus (63BC–AD14) as the *princeps* of Rome. Augustus, the first Emperor, restored peace and order in the Roman territories following six decades of civil wars that almost destroyed the political power of Rome. When Augustus became the sole ruler of the Roman territories, he faced an economy that had been shattered by the terrible military events of the previous years; a society divided by contrasting interests; and an overpopulated military employing many professional soldiers, many of whom were now at risk of being made unemployed.

Rome's political power had been damaged by civil conflicts: several peripheral territories were in revolt, some areas still had to be pacified, and the system of 'client kings' needed to be re-structured. The traditional Roman virtues of valour and honour, which had guided senators and generals of the Roman Republic during previous centuries, seemed to have been lost. Rome needed a restoration and to adopt a new form of government. The civil wars had shown that the Roman territories were extended too far to be governed by an assembly such as the Senate, which was made up of politicians with contrasting political interests. The institutions of the Republic had worked well when Rome controlled just the Italian peninsula, but now that the whole Mediterranean was under Roman control, a more effective and centralised form of government was needed.

Each of the territories that made up the late Roman Republic had a different institutional status: some were provinces, others were client states ruled by vassal kings, and there were also allied states with a higher degree of autonomy. This organisational chaos had to be brought to order for there to be more cohesion in the Roman territories. During the last civil war, Octavianus Augustus had defeated Mark Antony (83–30BC) and had added Egypt to the Roman territories. Egypt was not ruled as a standard province, but was attributed special status and placed under direct control of the princeps. As a result, the occupation of Egypt augmented the institutional fragmentation that characterised the Roman territories during the late years of the Republic.

To solve all the problems, Augustus transformed the Roman Republic into an Empire ruled by a single man. He knew, however, that such an important political change would never have been accepted with ease by the Senate and its occupants, the Roman oligarchs. Just 15 years before, his mentor, Julius Caesar (100–44BC), had tried to do something similar, and had been assassinated in the Senate by the Roman aristocracy. Augustus had to move very cautiously to carry out his planned reform without provoking strong opposition. Having the backing of common citizens, who were tired of civil wars and wanted peace, he gradually removed the most important powers from the Senate but retained the aristocrats' privileges.

The Senate continued to function, the republican institutions were preserved and – it appeared – the form of government remained unchanged. Augustus assumed the title of princeps, which is different from that of Emperor. He did not have absolute powers, and he always respected the prerogatives of the Senate. He defined himself as a *primum inter pares* (first among peers), in order to show his respect for the aristocracy. The Senate, understanding that a collaboration with the princeps could have only positive consequences, accepted the reforms with enough enthusiasm and did not oppose Rome's transition from Republic to Empire. The transition was completed after a few years and without any significant problems. As time progressed, Augustus became increasingly popular and started to transform the new government – known as a *principate* – into a form of hereditary monarchy. When he died, in AD14, the Senate had already become a purely ornamental institution, since all powers was now vested in the hands of the Emperor: the transition had been smooth and effective, the Empire was effectively born.

From AD18 until AD180 (the Early Empire), the successors of the first Emperor ruled Rome as absolute monarchs and expanded their territories by launching victorious military campaigns.

During the years AD180–284 (the Middle Empire), the Roman Army faced the terrible 'Crisis of the Third Century', which was the result of many different factors, including gigantic socio-economic problems, as well as the arrival of new Germanic populations on the borders of the Roman Empire. The Roman state, after the flourishing years of the Early Empire, risked being wiped out. Several important provinces became temporarily independent, and large areas of the Empire were raided by foreign invaders. The Roman Army had to fight to reconquer the Empire and to preserve its unity. To survive, the Roman military forces needed to modernise and reform, to face new threats posed by a multitude of war-like enemies, including the Persians in the Middle East, or the Germanic peoples in Central Europe. The old military system, based on the power of the legions and their heavy infantry was completely remoulded. Cavalry became much more important, as did light troops specialised in skirmishing. The introduction of new weapons permitted the development of innovative tactics.

In the book, we will follow the evolution of the Roman Army from the death of Marcus Aurelius (AD121–180) to the rise of Diocletian (AD242/254–311/312), reconstructing the major military campaigns fought during AD180–284 and explaining how the Roman military forces entered the Late Empire (AD284–476).

Chapter 1
The Roman Army of the Early Empire

Since the beginning of his rule as princeps, Augustus presented himself as the sincerest defender of peace. He understood that Roman society needed a period of peace and prosperity after having been shattered by terrible internal divisions. Internal order was restored by cleaning up the countryside from the bands of brigands that ravaged it, as well as by dissolving all the political associations associated with the defunct leaders of the civil conflicts.

Augustus paid particular attention to improving the living conditions of the poorest social groups. He reduced the number of slaves in society and re-categorised their living, so that each had duties but also some rights. He cancelled all debts contracted by the poorest citizens during the civil wars and reorganised the local administrations by introducing *equites* (middle-class men) into them. Augustus promoted demographic expansion with the promulgation of some specific laws. He regulated the monetary market to initiate a period of economic growth; and distributed the lands confiscated during the civil conflicts to his veteran soldiers, or to farmers of proven loyalty. He restored the traditional religion and morality of Rome after decades of bloody crime, and ordered the construction of many important public infrastructures, most notably of roads and aqueducts. An ambitious cultural programme was founded, which was functional to his political propaganda and was based on the importance of literature (it was under Augustus, for example, that Virgil wrote the *Aeneid* epic-celebrative poem). All these measures had a deep impact on the lives of ordinary citizens and soon brought Roman society and its economy into a period of great renaissance.

Augustus also had to manage the defence of the borders of his domains, which required reorganising. In 27BC, all the regions of the Empire, except for the Italian peninsula, which had special status, were reorganised as provinces. Provinces could be either *provinciae pacate* (peaceful) and *provinciae non pacate*.

The former consisted of territories conquered long ago and considered fully pacified and they were not exposed to serious military threats. The peaceful provinciae pacate were placed under the direct jurisdiction of Augustus, together with the large military contingents that garrisoned them. The provinces were governed by officials known as *legati Augusti pro praetore*, who were extremely loyal to Augustus and usually had solid military skills. The taxes collected in the provinciae pacate became part of the *aerarium* (treasury) of the Roman state, which was controlled by the Senate.

The provinciae non pacate status was allocated to those territories in which some form of local opposition to the Roman presence still existed, or that were particularly exposed to foreign attacks. These regions were managed by *proconsuls* (governors), who were aristocrats chosen by the Senate. The taxes collected here became part of the new *fiscus* (treasury) of the princeps created by Augustus. The provinces under control of the Senate were very few but quite rich; they comprised what is now southern Spain, the coastline of Tunisia and Algeria, the western half of Anatolia, Greece, Macedonia and southern France.

Augustus progressively reduced the number of 'client kingdoms' (vassals of the Roman Empire) that were located on the borders of the Roman territories by absorbing them into the Empire. In addition, he ordered the foundation of new Roman colonies in the various provinces with the objective of augmenting the number of Italic individuals (inhabitants of ancient Italy) who lived in the peripheral areas of the Empire. The princeps formed a very strong alliance with the equites, who became the backbone of his new Imperial administration; these men now had the opportunity of a brilliant career in the provinces thanks to their technical skills and loyalty to the Emperor, something that had been impossible previously. Power and wealth were now distributed in a more egalitarian way, which greatly increased the Emperor's popularity.

The reforms of Gaius Marius

When Augustus became princeps, the Roman Army was a deadly military machine comprised of professional soldiers. Following the Battle of Arausio in 105BC, the great general and consul, Gaius Marius (157–86BC), had reformed the military into a force made up of professional fighters, coming from the poorest social groups. Many recruits would have served in the army for most of their lives and would have been equipped by the state. These new professional soldiers would have earned a living in the military and would have been available to serve in every corner of the Mediterranean for as long as required. Gaius Marius greatly improved the training and discipline of the Roman military forces, giving each soldier his own personal equipment that included several working tools. As a result, each *legionary* (heavy infantryman) would have been able to operate as a combat engineer, if needed, to build fortified encampments or to destroy the fortifications of the enemy. Marius created a new *esprit de corps* (common spirit) inside the legions: he was the first, for example, to give a distinctive standard and a peculiar denomination to each unit of the Roman Army.

With the outbreak of the civil wars, the military reform championed by Gaius Marius was tested to the limit: opposing factions recruited armies with thousands and thousands of professional legionaries, which destroyed each other in some of the bloodiest battles of Antiquity. Yet, from a qualitative point of view, no army in the Mediterranean world could face the new Roman legions on equal terms: the Romans were better disciplined, trained, equipped and commanded than any of their potential opponents.

Reorganising the military

When the civil wars came to an end in 30BC, it became clear that the immense military machine that had been mobilised during previous decades had to be greatly reduced. The finances of Rome could no longer sustain the costs of keeping an army with 60 legions incorporating 300,000 men. In addition to the regular army, during the last years of the civil wars, several new legions had been formed with soldiers of inferior quality: sometimes every able-bodied man of a territory had been forced to enlist as a legionary, or entire armies of client kings had been transformed into legions. Augustus had to reduce the number of legions drastically and make a rigid selection of the best legionaries to remain in service. The new army was to consist of recruits who joined voluntarily because they wished to serve the Empire as professional soldiers for two decades of their lives; these would have been loyal to the Emperor more than to any other institution.

In order to be sure of his soldiers' loyalty, Augustus created the *aerarium militare* (military treasury), which was specifically tasked with paying the legionaries precisely and regularly. The Emperor knew that the only way to avoid mutinies and rebellions was to pay soldiers well (he increased their standard pay) and to offer them acceptable conditions. Once a legionary completed his service, he was given a farm by the state to compensate him for his 20 years of honoured service. As a result, the new legions started to consist of determined and disciplined men who believed that loyalty towards the Emperor could guarantee them success in their personal life.

To find the economic resources needed to create the military treasury, Augustus reduced the number of legions from 60 to 28. Only the best were retained in service. To avoid the outbreak of revolts, however,

A Roman legionary of the Early Empire, with *lorica hamata* armour. (Legio VI Victrix Cohors II Cimbria)

Augustus gave donations and land to those veterans discharged from the army (including to those who had served in Mark Antony's legions). During the period of the civil wars, the survival of a Roman soldier depended on the decisions of his commander, from whom he received his pay as well as land at the end of his career. Under the new regime, the legionaries remained professional soldiers but were now paid by the central administration of the state; and the farms given to the veterans were provided by the imperial administration, and thus the loyalty of the soldiers started to shift from their commanders to the ruling imperial dynasty.

Legions

The main military unit of the new Roman Army was the legion: a large heavy infantry formation, entirely composed of soldiers who had chosen to sign up and who held Roman citizenship. Each legion included approximately 5,500 men, divided into one elite *Cohors Prima* (first cohort), and nine standard line cohorts. The Cohors Prima, formed of the best soldiers of the legion, included 800 men arranged into five centuries, each containing 160 men. The other nine cohorts each had 480 legionaries in their number, each made up of six *centuria* (centuries) of 80 legionaries. Every century was composed of ten smaller units, formed by seven soldiers plus a single non-commissioned officer (NCO) known as *decanus*, and two non-combatants. This kind of unit, known as *contubernium*, was the smallest one existing in the Roman Army.

In addition, each legion had its own cavalry component: 120 cavalrymen divided into four *turmae*, each with 30 soldiers. Each *turmae* had three *decuriae*, the smallest cavalry units, which included eight soldiers, a *decurion*, the equivalent of the infantry centurion, and an *optio*, acting as non-commissioned office (NCO). The greater number of legionaries came from Italy, but as time progressed the provincial employees became increasingly numerous.

The system of ranks of the Roman legion was quite complex, since it was specifically designed to make such a large body of soldiers work efficiently. The overall commander of the legion was the *Legatus Legionis*, usually a senator appointed by the Emperor. As time progressed, however, many members of the equites social class became commanders of legions. In case a legion was garrisoned into a province where no other legions were located, its commander was to act as provincial governor. When performing these roles, he was commonly known as *Legatus Augusti pro praetore*. The second officer in the chain of command was the *Tribunus Laticlavius*: generally, he was a young member of the Senate with no military experience, who was assigned to the command of a legion in order to learn from the Legate. The third in command was the *Praefectus Castrorum*, who was responsible for the building and defence of the legion's camp. He was a long-serving veteran from the equites social class, who had already finished his 20 years of service and who had served as the most senior centurion in a legion. His functions included the training of soldiers and new recruits. The group of senior officers was completed by five lower-ranking tribunes known as *Tribuni Augusticlavii*. These all came from the equestrian class and already had military experience. They mainly performed administrative functions and commanded two cohorts each.

The most senior centurion in a legion was the *Primus Pilus* and commanded the first century of the elite First Cohort. Once in battle, he had the honour of commanding the entire Cohors Prima. The other five centurions of the First Cohort were known as *Primi Ordines* and were considered to be superior in rank to their equivalents of the other cohorts. The centurions commanding the first centuries of the nine cohorts were called *Pili Priores* and were superior in rank to the Primi Ordines of the First Cohort. Like the Primus Pilus, each of these officers took command of his entire cohort in case of battle. Generally, all centurions came from the lower social classes and became officers because of their personal capabilities: these experienced veterans were the backbone of the Roman Army.

Above left: A Roman legionary of the Early Empire, with lorica hamata armour. (Legio XIII Gemina)

Above right: A Roman legionary of the early Principate, with lorica hamata. The bronze discs and red leather straps were worn as military decorations only by the veteran soldiers. (Legio XI C.P.F. Hispaniensis)

The commander of the four cavalry turmae was known as *Tribunus Sexmentris*; the decurion commanding each first decuria of horsemen had the honour of leading the entire turma in battle. The rank organisation of the NCOs was particularly articulated, in order to perform a wide range of different functions. Each centurion directly appointed his adjutant and second in command, who was called *Optio*: this person performed more or less the same commanding functions of the centurion, but during battle he had the important responsibility of remaining at the back of the century in order to keep the formation closed and to discourage deserters from abandoning the field. Each Optio had an adjutant known as *Tesserarius*, who was third in the century's chain of command. The Tesserarius was keeper of the watchword of his unit and performed several important administrative functions. The system of NCOs was completed by the ten *Decani*, each of whom commanded one of the *contubernia* that made up a century. Two auxiliary servants were assigned to each contubernium.

Above left: A Roman legionary of the early Principate, with crested helmet. (Legio XIII Gemina)

Above right: A Roman legionary officer of the early Principate, with crested helmet and manica. (David Burns, Legio XIIII Gemina)

Common legionaries could perform a series of special duties, which had a specific designation and corresponded to higher pay. The most important of these was that of *Aquilifer*: the *Aquila* was the legion's standard and represented the valour of the whole unit; losing it was the greatest dishonour that a Roman legion could endure. As a result of this symbolic importance, its bearer was chosen among the most experienced veteran soldiers. Each cohort had its own *Vexillum* (standard) that was carried by a special bearer known as *Vexillifer*, while each century had its own *Signum* that was carried by a *Signifer*; this consisted of a spear shaft decorated with medallions and topped with an open hand to signify loyalty to the state. In addition, the Cohors Prima had the privilege of carrying a special image of the Emperor, which was given to an elite bearer known as *Imaginifer*.

Finally, each century had its own military music, consisting of a *Cornicen* or *Bucinator* and a *Tubicen*. All these transmitted orders by playing their instruments and usually acted in close combination with the Signifer, who was a rallying point for all the legionaries. The Cornicen played a horn, the Bucinator played a particular kind of horn known as *buccina* and the Tubicen played a long trumpet. Each legion deployed its own artillery: each cohort was responsible for employing a stone-throwing *ballista* and each century was responsible for employing a bolt-shooter *scorpio*. In total, a single Roman legion could deploy an impressive artillery detachment with ten heavy ballistae and 59 light *scorpiones*.

Light infantry and cavalry

With the military reform carried on by Gaius Marius during the Cimbrian War (113–101BC), the Roman legions adopted a new structure that did not comprise light infantry but *cohortes* of heavy infantry instead. As a result of this, the Romans relied on foreign soldiers during the late Republican period in order to deploy sizeable contingents of light troops and cavalry; the latter had never been a fundamental component of the Roman Army. During his Gallic campaigns, for example, Julius Caesar deployed the following auxiliary contingents: Celtic heavy cavalrymen, Germanic light cavalrymen, African mounted skirmishers recruited from the Numidians and the Mauri, Cretan archers, Balearic slingers and Iberian light infantrymen. The Celtic heavy cavalrymen came from the tribes of Gaul that were allied with Rome, most notably from the Aedui. They were equipped with helmet and chainmail and were armed with spears and long slashing swords. Mounted on tall horses, these cavalrymen made up a 'shock force' that was employed by Caesar on several occasions and with great success. The Germanic

A Roman veteran legionary of the early Principate, drinking from his canteen covered with wicker. The Celtic torque neck rings, worn as decorations, have probably been taken from defeated enemies. (Legio XIIII Gemina)

light cavalrymen, instead, were equipped as light skirmishers: they did not carry armour or helmets and were armed with throwing javelins. These horsemen were used to conduct reconnaissance missions as well as to harass the enemy with rapid incursions. On some occasions, they were also employed to launch frontal charges. Generally speaking, they were always considered by Caesar as a fundamental component of his military forces.

In 55bc, the Roman troops operating in Gaul comprised a total of 400 Germani, who were all nobles coming from the tribes that had been defeated by Caesar and who served alongside the Romans (together with their servants) as 'hostages'. Their presence on Rome's side was a sign of goodwill and trust, which had great political importance. Julius Caesar was particularly impressed by the martial spirit of the Germani: he admired them for their courage and for their simple way of life (which was ideal to 'forge' good warriors). The 400 Germanic cavalrymen who served as part of Caesar's army were supported by light infantrymen: these were the servants of the Germanic nobles and were trained to run together with the horsemen in order to cover their flanks during close combats. Caesar encountered this peculiar tactical formation, invented by the Germani, during the campaign fought against their leader Ariovistus (101–54bc), and soon understood that it had great potential. As a result, he was the first to create some combined units of infantry and cavalry inside the Roman Army; these later became known as *cohortes equitate* and remained an important component of the Roman military forces for a long time.

With the outbreak of Vercingetorix's (82–46bc) rebellion many 'allied' Celtic horsemen abandoned Caesar and thus the latter was obliged to recruit more Germani in order to replace them. These 600 soldiers were accompanied by their servants/light infantrymen. The Germanic warriors of Caesar fought with enormous valour on several occasions against Vercingetorix's forces and always routed the Celtic cavalry that tried to stop them. The Germani who participated in the final Gallic campaign of 52bc followed Caesar during the ensuing years, taking part in all the conflicts fought by their general and most notably in the civil war that ended with the defeat of Pompey at the Battle of Pharsalus (9 August 48bc).

In 45bc, Caesar returned to Rome and disbanded the auxiliary corps of his army; the surviving Germanic veterans went back to their homeland after having travelled around the Mediterranean and after having plundered in many different countries. The Numidians and the Mauri serving with Caesar were equipped as mounted skirmishers and used javelins as their main weapon; just as they had already done as part of Hannibal's Carthaginian Army. They performed as explorers and were masters at organising ambushes. Cretan archers and Balearic slingers were all mercenaries, who were well known in the Mediterranean world for their incredible fighting skills. Crete and the Balearic Islands were the home of two of the most important 'light infantry schools' of Antiquity, where the deadliest archers and slingers were trained as professional soldiers from childhood. The Iberian light infantrymen were known as *caetrati*, because they carried a small round shield that was called *caetra*. They were armed with deadly short swords and with javelins; Caesar appreciated them for their great tactical flexibility.

Auxiliaries

When Augustus reformed the structure of the Roman Army, he decided the foreign contingents of light infantrymen and cavalrymen described should become a fixed component of the Roman military forces. Up to that point, these light and mounted contingents had been recruited on a temporary basis and had always been disbanded at the end of military campaigns. They did not have a standardised organisation and could be commanded by their own officers, who were mercenaries paid by the Roman Republic. Augustus changed all this by transforming the provisional corps of light infantry and cavalry into a permanent component of the new Roman Army: the *auxilia* or auxiliaries. These would have been commanded by Roman officers and received regular pay from the imperial government; as a result, they would have lost their former mercenary status and become a proper military institution.

A Roman legionary of the Early Empire equipped with his working tools. (Legio XIII Gemina)

The new auxiliaries, like their predecessors, would have been recruited on a local basis from the provinces of the Empire; as a result, they would have provided contingents with different tactical features, coming from every corner of the Roman world. Different to the legionaries, they were not Roman citizens, but after 25 years of honoured military service they were to be granted Roman citizenship and a farm located in their home province. As a result of the new system introduced by Augustus, serving in the auxilia units became the best way for a provincial to become a Roman citizen and to improve his social position. The son of a provincial auxiliary could enlist in the Roman Army as a legionary. A new hereditary class of provincial soldiers was soon formed, with men who were strongly linked to their home provinces but at the same time were extremely loyal to Rome due to the opportunity for social improvement. As time progressed, the auxiliary units increased in professionalism and were sent out campaigning across the Empire. As a result, the outbreak of revolts in the auxiliary units supported by the population of their own provinces could be avoided.

With the expansion and consolidation of the Empire, new military traditions were gradually absorbed into the Roman Army thanks to the auxilia system, such as archers recruited from the eastern provinces, or camel troops. The Romans first encountered camel-mounted troops when they fought against the Seleucid Antiochus III in 190BC, but it was only during the second century AD that a Roman camel corps was formed. In AD106, Trajan added to the Empire the new province of Arabia Petraea and created the Ala I Ulpia Dromedariorum Palmyrenorum, which was formed with camel warriors coming from the city of Palmyra. In the following decades, the Romans formed various small auxilia units of *dromedarii*, because these proved to be very effective in patrolling the desert frontiers of the Empire. Their tasks included escorting convoys, defending important routes of communication, scouting in the desert, escorting couriers and battling desert bandits. They were organised in small squads, having special bases in the desert that provided them with food and water.

Auxiliary units

There were three different kinds of auxilia units: *cohortes* of all infantry, *alae* of all cavalry, and *cohortes equitate* (ie, infantry cohorts with an attached cavalry contingent). The cohorts had the same internal organisation and structure as the legionary armies, albeit acting as light infantry units. The alae were the proper cavalry of the Roman Army, since the few cavalrymen of the legions performed only auxiliary

Above left: **A Roman legionary with lorica segmentata and rectangular shield. (Marc Seriol (@marcmarkhus_photo), Legio II Traiana Fortis – Cohors I Barcinonum, Barcino Oriens)**

Above right: **A Roman legionary with lorica segmentata armour. (Legio VI Victrix Cohors II Cimbria)**

duties in function of the heavy infantry (such as escorting officers, transmitting orders or scouting). The alae could be either heavy or light cavalry, but were all characterised by their specific training. The cohortes equitate – created by Julius Caesar – were a mix of foot and mounted auxiliaries, a smaller replica of the larger legions: thanks to the presence of some horsemen they could act as autonomous small armies for service on the frontiers. Each of these three kinds of unit could have two different versions, with 500 or 1,000 men.

A *cohors quingenaria* had six centuries of 80 soldiers, with a *Praefectus* as commander. A *cohors miliaria* had ten centuries of 80 soldiers, with a *Tribunus Militum* as commander. The *Ala Quingenaria* had 16 turmae of 30 men, with a *Praefectus* as commander. The *Ala Miliaria* had 24 turmae of 30 men and a *Praefectus* as commander. A *Cohors Equitata Quingenaria* had six centuries of 80 infantrymen and four turmae of 30 cavalrymen, and were commanded by a Praefectus. A *Cohors Equitata Miliaria* had ten centuries of 80 infantrymen and eight turmae of 30 cavalrymen, and were commanded by a Tribunus Milituum. During military campaigns, the various auxiliary units were attached to the legions, thus coming under direct control of the latter's commanders.

Most of the auxilia military units were stationed on the vast borders of the Empire, together with the legions. The frontiers were the only points from which enemy attacks could come and so this defensive positioning of the Roman forces was logical. It should be pointed out, however, that in case of internal disturbances very few men were available to face threats coming from the inside of the Empire. There was no central military reserve and it was practically impossible to take a full legion from its frontier garrison without creating a major gap in the border defences. The only possible solution was of taking detachments from various legions and assembling them together in order to form temporary task forces known as *vexillationes*. As soon as the internal emergency was resolved, a temporary vexillatio was dissolved and its detachments were sent back to their parent legions. Usually a vexillatio was a mixed force of legionaries and auxiliaries, formed by 1,000 infantrymen and 500 horsemen: two infantry cohorts taken from a legion and a single ala quingenaria of mounted auxiliaries.

Praetorian Guard

The military condition of Italy and of Rome was quite different to that of the provinces: the capital was the heart of the Empire, where the Emperor and the central administration worked. According to the reforms of Sulla and to an ancient Roman tradition, no military units could enter the territory of the Italian peninsula without representing a menace for the institutions of Rome. Italy was thus considered to be sacred by the Romans, who referred to Rome as the *pomerium*. The Rubicon River in the north and the Straits of Messana in the south were Italy's natural borders. As a result, Italy could not be garrisoned by legions or by auxiliary units in the same way as the other provinces. Yet the Emperor and the city of Rome required some military protection (especially against internal threats). Augustus resolved this issue by creating the famous *Praetorian Guard*: this unit, formally the personal guard of the Emperor, was also to act as the garrison of the Italian peninsula. Initially, it was formed with the best soldiers from the legions, so it represented the elite of the Roman Army. Created by Augustus and later reorganised by his successor Tiberius (42BC–AD37), the Praetorian Guard was to be supported by a structure of nine infantry cohorts. Under the first Emperor, three cohorts were stationed in Rome and the other six in other important cities of Italy. Later, during the reign of Tiberius, all the Praetorians were garrisoned in Rome, in the famous barracks known as *Castra Praetoria*. The whole Praetorian Guard was commanded by a *Praefectus Praetorio*, who came from the ranks of the equites.

Each single cohort was commanded by a Tribunus Militum and had six centuries of 80 soldiers. The original number of nine cohorts, however, was later changed: around AD47 it increased to 12, and in AD69 it became 16 (under Vitellius (15–69AD), who also augmented their internal establishment from 500 to 1,000 men each). Vespasian (9–79AD) reduced the number of cohorts to nine (with 500 men each), while Domitian (51–96AD) increased it to ten, setting the definitive establishment that would not change until final dissolution of the corps. Initially, all the Praetorians were ex-legionaries coming from Italy, but over time the presence of provincials became increasingly important.

The soldiers of the Praetorian Guard had many privileges when compared with the standard legionaries: their period of service lasted for 16 years instead of 20 and their pay was significantly higher than that of standard soldiers. Trajan (53–117AD) added a cavalry component to the Praetorian Guard, whose members were known as *equites singulares Augusti*: these mounted Praetorians escorted the Emperor whenever he left Rome for a military campaign or for a journey in the provinces. The equites singulares Augusti were organised as a standard ala miliaria, with 24 turmae of 30 men and a Tribunus Militum as

Above left: A Roman legionary with lorica segmentata. (Legio XI C.P.F. Hispaniensis)

Above right: A Roman legionary with lorica segmentata and leather pteruges. (Legio XI C.P.F. Hispaniensis)

commander. All the members of this unit were selected from the best horsemen of the auxiliary alae and were granted Roman citizenship upon enlisting into the Praetorian Guard.

Urbaniciani and Vigiles

The military reform of Augustus that affected the garrison of Italy was not limited to the creation of the Praetorian Guard, since he formed another two corps that were to be stationed in Rome: the *Urbaniciani* and the *Vigiles*. The first were a sort of police, whose main function was to keep order in the immense city of Rome. If the Praetorians were the protectors of the Emperor, the Urbaniciani were the protectors of Rome. While the Praetorian Guard was under direct control of the Emperor, the Urbaniciani was the only Roman military unit still under control of the Senate: the commander, the *Praefectus Urbi*, was, in fact, a senator. Initially, the Urbaniciani consisted of three cohorts, numbered progressively after the nine Praetorians; each single cohort was commanded by a Tribunus Militum and had six centuries of 80 soldiers. The original three cohorts were later expanded to seven during the reign of Claudius (10BC–AD54), while Vitellius increased the size of each cohort to 1,000 men. Different to the Praetorians, the Urbaniciani continued to be recruited from Italy during their entire existence.

The Vigiles corps created by Augustus were stationed in Rome and acted as urban guards/firemen. Their main functions, in fact, were patrolling the streets during night and fighting fires (which could be destructive for a city of the time, as the Great Fire of Rome clearly showed in AD64). The commander was known as *Praefectus Vigilum* and came from the class of the equites. Vigiles were organised in seven cohortes miliariae for a total of 7,000 men. Most of their members were *liberti* (freed slaves) of low social condition. Each cohort of the Vigiles was commanded by a Tribunus Militum and was peculiarly divided into just seven centuries. At the time of Augustus, the city of Rome was divided into 14 regions, so each cohort of firemen was responsible for the security of two regions.

Private bodyguards

Finally, Augustus created a small corps of foreign bodyguards in order to protect his own person: the *Germani corporis custodes*. This unit was 'private' guard, totally independent from the Praetorians: it

was entirely composed of Germanic warriors of the Batavian tribe, who acted as personal protectors of the Emperor and of his family. Being foreigners, they were totally loyal to the monarch and had no links with the internal political situation of Rome. In AD9, as a result of the Roman defeat at the Battle of the Teutoburg Forest, this unit was temporarily disbanded; then reformed a few years later. In AD69, the Germani corporis custodes were definitively disbanded by Galba because of their strong loyalty to Nero. Initially, there were 100 Germanic guards, but by the time of Nero (37–68AD) there were 500; they were organised in decuriae, each of which was led by a decurion.

Roman centurion (left) and legionary (right), both wearing lorica segmentata. (Legio XIIII Gemina)

Chapter 2

The Military Campaigns of Augustus, 30BC–AD14

Soon after becoming Emperor, in 29BC, Augustus had to deal with problems at the frontiers of the Roman territories. Following their victories over Carthage during the Second Punic War, the Romans had occupied a large portion of the Iberian Peninsula and had established new provinces in that area of the Mediterranean world. The resistance of the local Iberian tribes to Roman penetration, however, was particularly strong. The Iberians were skilled warriors, who had been the backbone of Hannibal's victorious army. During the period 200–100BC, Rome consolidated its position in Iberia by expanding territorial possessions and by crushing several revolts of the local peoples. At the end of this long process, in which several Roman legionaries were defeated on the harsh terrain of present-day Spain, most of Iberia had been conquered by the Republic. Only the north-western corner of the Iberian Peninsula, consisting of Cantabria and Asturias, remained independent. This area was populated by the Celtiberians, an ethnic group of ancient Spain that was the result of the encounter between Celts and Iberian, Cantabria and Asturias. Being mostly covered with mountains, this area had a harsh climate and were quite isolated from the rest of Spain.

The outbreak of civil wars prevented the Romans from completing their definitive conquest of Iberia, since the legions spent most of their time fighting each other. The Spanish territories of Rome were involved in various conflicts on several occasions, for example, when the Roman general Sertorius (BC126–73) revolted against the Republic and allied himself with the Iberians (80–72BC), or when Pompey's (BC106–48) legions in Spain were defeated by Caesar's military forces at the Battle of Munda (45BC). After the end of the civil wars, Augustus decided to pacify Iberia in a definitive way by conquering Cantabria and Asturias. The Celtiberians living in the latter regions were strongly determined to preserve their independence. They lived between the cliffs of the Pyrenees and the waters of the Atlantic Ocean, in the wildest region of Spain. The Cantabrians had served as mercenaries for both the Carthaginians and the Romans during the previous centuries and had been appreciated as light infantrymen by Julius Caesar. In contrast, the Astures had always lived in a state of almost complete isolation. Both tribes used guerrilla tactics based on hit-and-run attacks and were experts in mountain warfare.

Iberia

The standard panoply of the Celtiberian warriors consisted of light weapons: short swords, daggers, short spears, throwing javelins, and small shields of round or oval shape (this model of shield was called caetra by the Romans). Knowing that the conquest of Cantabria and Asturias would have been a long and difficult process, the Emperor deployed substantial numbers of men against the Celtiberians. During the various campaigns that were fought by Augustus in northern Iberi, the following Roman military units contributed:

- I Legio *Augusta*
- II Legio *Augusta*
- IV Legio *Macedonica*
- V Legio *Alaudae*
- VI Legio *Victrix*
- IX Legio *Hispana*
- X Legio *Gemina*

- XX Legio *Valeria Victrix*
- Ala *Augusta*
- Ala *Parthorum*
- Ala II *Gallorum*
- Ala II *Thracum Victrix Civium Romanorum*
- Cohors II *Gallorum*
- Cohors IV *Thracum Aequitata*

Above left: A Roman legionary with lorica segmentata armour. (Legio VI Victrix Cohors II Cimbria)

Above right: A Roman veteran legionary with lorica segmentata and vexillum. (Legio XIIII Gemina)

The first years of the conflict, 29–26BC, saw little progress for the Romans who were unable to penetrate enemy territory. The elusive tactics of the Celtiberians, in fact, prevented them from obtaining a decisive victory on the open field. During the last months of 26BC, Augustus assumed overall command of his military forces in Spain and established a large base near the Esla River, where three legions were stationed. Over the following months, the Roman troops again attacked the Astures and gradually forced them to take refuge in their hillfort of Lancia, described as the most important stronghold of the Celtiberians by contemporary Roman historians. Before the legionaries could besiege them in Lancia, the Astures abandoned their fortification and made for a nearby mountain called Mons Medullius by the Romans. Here they built a large encampment and prepared new defensive positions. The Roman legions besieged the latter, building a 15-mile-long moat and ditch around the Celtiberian positions; clearly Augustus wanted to trap the enemies and to avoid the potential for the Astures to come down from the mountain to launch raids against Roman positions. Upon realising that their positions could not hold out for long due to their lack of food and other supplies, the Astures commited suicide en masse instead of surrendering. The Romans were shocked by such an act of desperate courage, which was the result of the Celtiberians' love for freedom.

Following the defeat of the Astures, the conflict continued for several more years, since the Cantabrians were determined to resist. The Romans experienced serious logistical difficulties and, more importantly, were never able to engage their enemies in a pitched battle. Augustus, who also had health problems during his stay in Spain, selected very harsh methods: the Celtiberian civilians were treated like enemy warriors and the territory of Cantabria was looted with no mercy. In order to avoid slavery, the Cantabrians would usually commit suicide when captured or surrounded. As a result, the whole war was extremely violent and atrocities were committed by both sides. After ten years of campaigning, in 19BC, the major fighting came to an end since all the Cantabrian communities had been defeated one by one. Sporadic rebellions took place until 16BC, but in practice, the Emperor had pacified northern Spain and had completed the conquest of Iberia. It should be noted, however, that two legions of the Roman Army (the X *Gemina* and the IV *Macedonica*) were left as a garrison in Asturias and Cantabria for seven decades due to the Romans' fear of new Celtiberian uprisings.

Illyricum, Dalmatia and Pannonia

The next war fought by Augustus as princeps was the *Bellum Batonianum* or Great Illyrian Revolt, which took place during AD6–9 in the regions of Illyricum, Dalmatia and Pannonia. The latter corresponded, more or less, to the modern nations of Slovenia, Croatia and Hungary. The Roman Republic had conquered the western Balkans with great difficulty and after having fought three bloody wars against the local tribes of the Illyrians, the Roman legions finally crushed them in a definitive way. The Illyrians were masters of the Adriatic Sea before the ascendancy of Rome and were among the most feared pirates of Antiquity together with the Cilicians. The main cause of the outbreak of the Illyrian Wars had been the Romans' will to destroy the naval power of the tribes living in the western Balkans. While the Republic achieved its main objective, but the Illyrian territories were never fully pacified by Rome. The conquerors initially ruled Illyricum as a protectorate, which was transformed into a provincia pacata during 27BC. Very soon, however, Augustus understood that Illyricum still needed a massive Roman military presence due to the possible outbreak of internal revolts. As a result, the region was transformed into a provincia non pacata. Before becoming Emperor, as a young military commander, Augustus had conducted a campaign in Illyricum; the latter, in fact, had lived in a state of complete anarchy during most of the previous decades due to the ongoing Roman civil wars. The Illyrian pirates, never completely subdued by Rome, had taken advantage of their master's temporary weakness to conduct new naval raids across the Adriatic Sea and to attack the Roman

A Roman legionary with lorica segmentata and leather pteruges. (Legio XI C.P.F. Hispaniensis)

military garrisons in the northern Balkans. Order was temporarily restored by the Romans before the outbreak of the civil war between Augustus and Mark Anthony, but the whole area always remained on the verge of revolt.

To have more direct control over the coastal areas where the potential pirates lived, Augustus separated the coastal region of the western Balkans from the interior region: the former became the new province of Dalmatia, while the latter continued to be known as the province of Illyricum. During 14–10BC, there were a series of minor rebellions in northern Dalmatia, during which the local tribes were supported by the peoples of Pannonia. The latter region was still independent from Rome and was inhabited by a mix of Illyrian and Celtic tribes; these included some Germanic elements, who were very warlike and extremely hostile to Rome. After more years of skirmishes and minor campaigns, the peoples of Illyria and Dalmatia finally joined forces with those of the Pannonians in order to launch a large-scale insurrection against Rome.

The following Illyrian tribes from Illyricum and Dalmatia formed an anti-Roman military alliance: the Daesitiatae, the Breuci, the Dalmatae, the Andizetes, the Pirustae, the Liburnians and the Iapydes. Both the Dalmatae and the Breuci were commanded by a warlord named Bato, and thus the new conflict became known as Bellum Batonianum. It was extremely difficult for the Romans from the beginning, because the revolting tribes had an efficient military organisation that was partly modelled on the Roman machine: thousands of the rebelling warriors had previously served in the Roman auxilia units. Like the Celtiberians, the Illyrians and the Dalmatians always employed hit-and-run tactics against the Romans and avoided pitched battles. As a result, most of the Great Illyrian Revolt consisted of counter-insurgency operations carried on – with some difficulties – by the Romans.

In AD6, Tiberius, the future successor of Augustus, asked the governor of Illyricum to join him in Germany with his military forces and to raise more auxiliary units from the Illyrians. The latter, once gathered, rebelled against the Romans and expelled them from their home province. Taking advantage of the Roman military weakness in that sector of the Empire, the rebels organised three different armies: one was to invade Italy, one was to invade Macedonia and the third would have remained in the western Balkans to prevent a future Roman counter-offensive. Augustus, greatly alarmed by these events, soon started to gather substantial military forces in order to stop an eventual Illyrian attack against Italy.

Meanwhile, in Macedonia, the rebels obtained a series of successes until they were stopped and defeated by the Thracians of the Odrysian Kingdom (a vassal state of the Empire). Once enough soldiers were assembled, Tiberius marched on Illyricum with the objective of crushing the large revolt. The future monarch, however, had a passive demeanour and was not able to engage the enemy in any major battle. The war of attrition continued during AD7, until the great military commander Germanicus (15BC–AD19) was sent to help Tiberius with substantial reinforcements. The Romans now had three legions in the western Balkans, together with many auxiliary units and together pushed back the rebels to the Claudian Mountains, located between northern Dalmatia and southern Pannonia, where a pitched battle was finally fought and won by Germanicus.

At this point, the Roman forces were divided into several detachments with the objective of chasing up bands of insurgents that were still active. The legionaries used very harsh repressive methods during this phase of the conflict, putting in practice the 'scorched earth' strategy that had been sponsored by Tiberius. By AD8, the civilians of both Dalmatia and Pannonia, ravaged by famine and by diseases, were ready to surrender, but were prevented from doing so by the rebel warlords who wanted to continue fighting. At some point the Dalmatians and the Pannonians turned against each other, and the latter surrendered. In AD9, the Romans reconquered the whole of Dalmatia and defeated the remaining rebels in a decisive way. With the end of the Great Illyrian Revolt, both Illyricum and Dalmatia were fully pacified. The Romans extended their sphere of influence over Pannonia. The latter was gradually occupied during the following years and was transformed into a new province during AD14.

A Roman legionary with lorica segmentata and decorative bronze discs. (David Burns, Legio XIIII Gemina)

Gaul and the Germani

Back in 44BC, by the time of Julius Caesar's assassination in the Senate, Gaul had been conquered by the Roman Republic. The Republic now had a very long border in common with the lands inhabited by the Germani, collectively called Germania by the Romans. Until 30BC, the military situation remained extremely quiet on the Rhine frontier, since the Romans were heavily involved in the civil wars that followed the death of Caesar. These ended only with the definitive victory of Augustus, who emerged as the first Emperor of Rome and who gradually transformed the Republic into a Principate. During the years that saw the consolidation of Augustus's personal power, several of the Celtic tribes living in Gaul revolted and tried to restore their previous autonomy. These local rebellions were all easily crushed by the Romans and by 27BC peace reigned again in Gaul. Augustus, however, was greatly irritated by these revolts and that the people had received material and ideological support

A Roman legionary wearing winter clothes of Celtic fashion under his armour. (Legio XI C.P.F. Hispaniensis)

from Germanic tribes living on the eastern bank of the Rhine.

As a result, to secure the Roman military presence in Rhineland, he built new fortifications in the area and severely punished the Germani who had helped the Gallic rebels. Augustus was gradually developing the idea of *limes* (fortified borders). However, the Roman legions could have easily conquered most of Germania, thanks to their military superiority. The Emperor was sure that the Germani could have been tamed easily because of the cultural superiority of the Roman civilisation. In his view, they had to be transformed from wild 'barbarians' to civilised subjects of Empire, as the Romans had already done with most of the Celts living in Gaul.

During the long reign of Augustus, the Romans absorbed several small client kingdoms into their Empire by forming a social alliance with local aristocracies. The nobles of the various protectorates were permitted to continue ruling their countries but were required to send their sons to the capital of the Empire to receive a perfect Roman education. Upon returning to their homeland, these 'client princes' would have been obliged to favour the incorporation of their homeland into the Empire and then governed it as a province for Rome. Augustus wanted to use this method of indirect conquest for Germania.

In 16BC, the Romans were given the *casus belli* that they needed to initiate military operations in Germania. Three tribes living on the eastern bank of the Rhine, crossed the frontier and attacked the Roman military forces garrisoning the border. The V Legion of the Roman Army was taken by surprise and defeated by the Germanic raiders, who were able to capture its eagle (the most sacred insigna of a legion). This unexpected defeat convinced Augustus to transform Gaul into an immense military camp and to begin preparations for an invasion of Germania. The recent events had shown that Gaul could only be considered completely safe after the elimination of the potential menace represented by the Germani. The tribes that had been responsible for the raids and the Roman defeat were pushed back on the eastern bank of the Rhine and sent peace requests to Augustus. He, however, was ready to implement his expansionistic policy towards Germania.

Taxes were collected from every corner of Gaul to finance the upcoming military expeditions and a mint was established at Lugdunum (modern Lyon) in order to coin the money that was needed to pay the legionaries for their work.

Drusus (38–9BC), a general with great combat experience and a stepson of Augustus, was made governor of Gaul in 13BC. According to Roman plans, he would have been the main commander of the military forces tasked with invading Germania. Soon after his appointment, however, Drusus faced a series of minor rebellions caused by the new taxation recently imposed by the Romans. While crushing the Gallic revolts, Drusus also had to repulse a Germanic raid from the eastern bank of the Rhine. He responded to this by launching a retaliatory attack across the river, which caused significant losses to the Germani.

Above left: A Roman legionary with woolen cloak worn during cold months. (Legio XIIII Gemina)

Above right: A Roman legionary with lorica segmentata and leather pteruges. (Legio XI C.P.F. Hispaniensis)

During 12–11bc, Drusus operated in north-western Germany with great success. He invaded the land of the Usipetes and pillaged the territory of the Sicambri, two of the tribes that had attacked Gaul. He then moved to the territory of present day Netherlands, where the local Germanic communities of the Batavi and the Frisians were easily subjugated. The latter soon became loyal 'clients' of Rome, since they provided soldiers for the Roman auxilia military units and even sent some chosen warriors who made up Augustus's private bodyguard.

After securing Roman dominance over the Batavi and the Frisians, Drusus moved along the northern coast of Germania and attacked the Chauci. Following other minor successes, he returned to Gaul due to the onset of winter. In a few months, without experiencing significant military difficulties, the Romans had occupied the actual territory of the Netherlands and had punished the most important Germanic tribes living near the Rhine for their incursions against Gaul.

During the following spring, Drusus launched another campaign in Germania. He rapidly subdued the Usipetes and then marched east in order to reach the Weser River. At this point, the Romans were already in the heart of Germania, where the important tribe of the Cherusci lived. The Cherusci controlled a large portion of territory, which was located between the rivers Ems and Elbe. No Roman general had ever advanced so deeply into Germania and thus Drusus was moving into unknown land. With winter approaching again and not being completely sure of the subdued Germani's loyalty towards the Empire, Drusus returned to Gaul. During the retreat, the Romans were attacked by small parties of Germanic warriors who harassed their columns by using hit-and-run tactics. The latter caused significant losses to the legionaries, who were obliged to move very slowly on a terrain that was mostly unknown to them.

At that time, the Rhine marked a 'frontier of civilisations' between Gaul and Germania. In Gaul, there were roads and *oppida* (urban settlements) built by the Celts, while in Germania the population lived in isolated settlements and no roads existed. Most of the Germanic lands were covered by impenetrable forests, characterised by a very humid climate and the presence of swamps. The heavy-armoured legionaries always had difficulty when moving across Germania, since their slow columns and their heavy equipment needed properly constructed roads to move rapidly. In the spring of 10bc, after having been elected Consul, Drusus, with his troops, crossed the Rhine again into Germania. This time, he concentrated his efforts against the important tribe of the Chatti, who lived south of the previously subdued Germanic communities. The Romans were able to subdue the Germani and gain local victories, but the new success of Drusus was not a decisive one. At the end of each Germanic campaign, in fact, the legionaries retreated to Gaul without leaving any garrison behind that could control the newly submitted territories. In practice, the military campaigns of Drusus were just demonstrations aimed at transforming the Germanic leaders into vassal kings and not invasions aimed at conquering new provinces in a stable way. In fact, the logistical difficulties experienced by the Romans prevented them from establishing a fixed military presence in Germania. All the gains obtained during spring and summer were lost during the cold months and so the Germani could not be considered as 'submitted'.

In 9bc, Drusus began his fourth Germanic campaign, again attacking the Chatti. This new campaign was particularly difficult for Drusus and his military forces, since they had to stop and fight several times during their march. The Romans again reached the land of the Cherusci and mounted a new attack on them. This campaign was more massive than the previous one and obliged the Germanic tribe to fall back to the Elbe River. The legions crossed most of the large territory inhabited by the Cherusci and devastated it, pillaging and looting with great violence. Drusus had understood that the Cherusci was the strongest Germanic community living in the central part of Germania and thus, by crushing their spirit, he hoped to gain the respect of all the other tribes. With the Romans in close pursuit and running the risk of being massacred, the Cherusci had no choice but to accept Drusus. They accepted formally the suzerainty of Rome over their homeland. However, the Romans were obliged to leave Germania with the coming of winter again and thus their supremacy over the lands of the Cherusci was not a stable one.

During his four Germanic campaigns, Drusus had been able to obtain several victories over the Germani: these, however, were more the result of his enemy's internal divisions than of his own military ability. The various Germanic tribes, in fact, were extremely fragmented politically and spent most of their energies in bloody inter-tribal conflicts. As a result, the Romans could create alliances with single tribes and use the local rivalries to their advantage, attacking the various Germanic communities one by one. The Romans always had numerical superiority over their enemy and could always prevail because of the assured discipline of their legions. The campaigns of Drusus, however, had also shown all the dangers that could be encountered in Germania: the latter was an unknown and savage land, where the Roman Army could not fight with its usual tactics and where the terrain was an ally of the local populations. Drusus had been intelligent enough to follow Caesar's example and had used the tribal divisions of his opponents to obtain some victories, according to the famous principle of *divide et impera* (divide and conquer). Due to logistical difficulties, however, he had been obliged to abandon his conquest every year with the coming of winter and thus the Roman presence in Germania remained a temporary one. On his way back to the Rhine, at the end of his fourth Germanic campaign, Drusus fell from his horse and was badly wounded. In a very short time, his injury became seriously infected and a month later, the great Roman general died of gangrene.

A Roman legionary wearing the manic" and greaves introduced during the Dacian Wars. (Legio XI C.P.F. Hispaniensis)

Knowing that Drusus was seriously ill, Augustus sent Tiberius to Germania as new military leader of the Roman military forces operating across the Rhine. Tiberius, brother of Drusus, did not have military experience but was extremely loyal to the Emperor: according to the latter's opinion, he would have soon learned from experience. During 8–7BC, Tiberius fought two campaigns in Germania, continuing what his brother had started. The efforts of the future Emperor were mostly directed against the Sicambri, a tribe that put up strong resistance. Tiberius had to use all his military resources to defeat these Germani, who used guerrilla tactics to slow down the advances of the legionaries and to cause them significant losses. The Romans responded by using very harsh, repressive methods, which included mass exterminations and the deportation of many Sicambri on the western bank of the Rhine. Tiberius, in fact, was sure that these Germani would have continued to revolt forever if the Romans had permitted them to remain in their homeland. After the end of Tiberius's second Germanic campaign, the Roman propaganda presented western Germania as a pacified region of the Empire. This, however, was not true as the subsequent military events taking place on the eastern bank of the Rhine would show.

A Roman legionary with lorica segmentata. (Legio XI C.P.F. Hispaniensis)

In reality, most of the Germanic tribes who had been defeated by the legions had accepted Roman supremacy only notionally and thus continued to act as autonomous communities. In addition, the Romans had no fortifications and no significant garrisons in Germania that could control the territory and prevent the outbreak of rebellions. Some of the most important Germanic tribes, living in the eastern part of Germania, had not been reached by the Romans and thus most of central Europe was still completely free from the political influence of the Empire.

In 6BC, Tiberius returned to Rome and was substituted as Roman military commander on the Rhine by Lucius Domitius Ahenobarbus (98–48BC). The latter resumed the Roman offensives in Germania and rapidly moved east, crossing the Elbe River. Ahenobarbus was a skilled military commander and understood the importance of building effective infrastructure in order to obtain stable control over the Germanic territories. He ordered his military engineers to build causeways across all the bogs that could be found in the region between the Rhine and the Ems so that the Roman legions could reach the Ems River much more rapidly than before. The building of these *pontes longi*, which would have soon after be connected by a road, was perceived as a clear act of hostility by the Cherusci who lived in the area.

After establishing permanent infrastructures, in fact, the Romans would have been able to deploy strong military garrisons in the region. As a result, in 2BC, the Cherusci rose up in revolt against the Romans and started to attack the legionaries that patrolled their homeland. It should be noted, however, that not all the members of this Germanic tribe were in favour of fighting a new war against the Empire: most of the common warriors hated the Romans and hoped to expel them from their territories. In contrast, a portion of the aristocracy was becoming increasingly interested in the new commercial possibilities offered by the arrival of the newcomers. The Rhine, after Julius Caesar's conquest of Gaul, was not perceived as a real barrier by most of the peoples living on its shores: the great river, in fact, was a waterway that connected different civilisations and could be used to transport commercial products. The Romans had been trading with the Germani long before the ascendancy of Augustus, and now that they had a stronger presence on the western borders of Germania, many wealthy aristocrats of the Germanic tribes understood that they could become rich traders. Some products that could be found only in Germania, such as amber, were particularly appealing to the Romans; likewise, other goods such as wine, which were not produced in Germania, were loved by the Germani. Regardless, the great majority of the Cherusci was in favour of war against Rome and so hostilities commenced. Even after many minor battles, Ahenobarbus did not obtain any significant victory over the Cherusci and on this account was rapidly substituted by the Emperor. In 2BC, in fact, Augustus sent a new military commander to Germania: the seasoned Marcus Vinicius (5BC–AD46).

Vinicius operated in Germania until AD4, at the head of a considerable military force that comprised five legions. He spent most of his years as commander of the Roman troops in Germania quelling local rebellions and marching across the Rhine to consolidate the Imperial presence on the eastern bank of the great river. He obtained several victories, but none was truly decisive since the Germani always avoided facing the Romans in a large pitched clash. In AD4, Augustus again sent Tiberius to the Rhine, with orders to pacify the region in a decisive way. The future Emperor performed extremely well during his second Germanic campaign, being able to subdue several minor tribes as well as to defeat the Cherusci.

According to Tiberius's propaganda, the Cherusci had been completely pacified and so were given the new and honorific title of 'friends of the Roman People'. In reality, Tiberius established a formal alliance only with part of the tribe's nobility and not with the great majority of its warriors (who still hated the Romans and their culture). In AD5, the Romans organised a major offensive in Germania, which began with a massive attack on the Chauci and continued with a rapid advance directed towards the heart of the Germanic territories. The Roman troops advanced on land, but a portion were transported on a newly built river fleet. These units met on the banks of the Elbe, where Tiberius assembled them into a single

Above left: A Roman legionary with crested helmet and lorica segmentata. (David Burns, Legio XIIII Gemina)

Above right: A Roman legionary with bronze helmet and oval shield. (David Burns, Legio XIIII Gemina)

army and made a strong demonstration of force in front of the Germani. After this, the Roman military forces returned to their starting positions without leaving any garrison in Germania. Once again, the Roman Army had penetrated into the Germanic territories with a rapid offensive, but it had been unable to defeat the Germani in a pitched battle. During the retreat, as before, the Romans were attacked by the Germanic warriors with hit-and-run tactics.

Battle of the Teutoburg Forest

The only positive result of the Roman demonstration of AD5 was that the Cherusci, after years of internal divisions and of minor conflicts with the Romans, finally decided to become bonafide allies of Rome.

A Roman legionary with scale armour. (Legio VI Victrix Cohors II Cimbria)

This change of political attitude was the result of an internal political struggle between the nobles of the Cherusci who wanted to establish a formal alliance with Rome and those who didn't. The rapid campaign of Tiberius convinced the majority of the Cherusci that, at least for the moment, the Romans were too strong to be faced in a proper military confrontation. In addition, the Germani had understood that by trading with the Romans they could improve their general living conditions and introduce new products to their economy. The pro-Roman party of the Cherusci tribe was guided by a powerful familiar clan, whose members included a young aristocrat named Arminius (d.AD21). Arminius was a sort of prince, since his father was one of the most important leaders of the Cherusci and a significant supporter of the new pro-Roman policy.

Tiberius showed support to the pro-Roman Cherusci and Arminius's clan in every possible way, in the hope of transforming the Cherusci into Rome's most important Germanic ally. To sustain his local allies and to keep an eye on the Cherusci who still resented Roman rule, he built a stable winter base on the Lippe River. For the first time, the Romans had a sizeable garrison in Germania. By AD6, the Romans exerted some form of direct or indirect influence over the entire western part of Germania, and all the tribes living west of the Elbe River recognised (at least formally) the supremacy of Rome. In south-eastern Germany, west of the Elbe River, there was just one Germanic tribe that had not yet been subjugated by the Romans – the Marcomanni, guided by their 'king' Maroboduus (30BC–AD37), and the strongest of all the Germani militarily.

The Romans, being sure of having complete control over all the other Germanic communities, planned a massive pincer attack on the Marcomanni that would have involved an impressive 12 legions. Before the latter could materialise, however, the Roman Army had to face the Great Illyrian Revolt, and thus the military forces assembled for the campaign against the Marcomanni were redirected to the province of

Illyricum. As a result, the Empire concluded a temporary peace treaty with the Marcomanni and recognised Maroboduus as their legitimate leader. Tiberius, who had guided the Roman troops with great intelligence in Germania, assumed command of the military forces fighting in the western Balkans and thus a new commander had to be chosen by Augustus in order to replace him: Publius Quintilius Varus (d.AD9).

At this point, it is important to understand who Arminius and Varus were. The Germanic prince Arminius, was born in 18 or 17BC, the son of a powerful noble of the Cherusci tribe. Following Drusus's campaign against the Cherusci of 9BC, he was sent to Rome as a hostage. Augustus had, in fact, been determined on 'cultural conquest' of the Cherusci, a strategy employed by the Romans with other client princes. By educating the leaders of their opposition and showing them the superiority of the Roman civilisation, the Emperor hoped to obtain a peaceful submission of their home territories after their princes returned to their homeland to rule as vassals of Rome. Arminius was raised in Rome and was educated as a perfect Roman boy: he entered the ranks of the army at a very young age and learned about Roman culture. Since the beginning of his military career, Arminius was blessed with great charisma and

A Roman legionary with leather manica, bronze greaves and scale armour. (David Burns, Legio XIIII Gemina)

impressive personal abilities. In reward for his military services, he was granted Roman citizenship and was made part of the equites social class (inferior only to that of the senatorial aristocracy).

During the Great Illyrian Revolt that broke out in AD6 he served under Tiberius, commanding a unit of mounted auxilia made up of Germanic soldiers. Augustus was impressed by the loyalty and combat skills of Arminius, who fought with distinction on several occasions. As a result, the Emperor ordered the Germanic prince to return to his homeland and to act as the personal 'aid' of the new governor Publius Quintilius Varus.

A Roman legionary with his full complement of personal equipment. (Marc Seriol (@marcmarkhus_photo), Legio II Traiana Fortis – Cohors I Barcinonum, Barcino Oriens)

Varus came from a senatorial family in Rome and had been Consul in 13BC together with Tiberius, who was a personal friend. He had married Vipsania (36BC–AD20), a daughter of Agrippa (63–12BC), who was Augustus's most trusted general and greatest personal friend. With such connections, Varus had a brilliant career in the Roman administration despite being considered inept by many. During the years 7–4BC, he was the Roman governor of Syria, where his personal limits became clear. He had no respect for the traditions of the peoples living on the territory of his province and his response to all problems was harsh repression. Additionally, he increased taxes in a significant way, and augmented his personal fortune illegally. In 4BC, soon after the death of the local Roman client king Herod the Great (b.72BC), a popular revolt broke out in Judaea. Varus was forced to intervene leading the four legions that were under his command in Syria. After occupying Jerusalem, he crushed the rebels with a violence never witnessed in Judaea. Without any formal process, Varus crucified 2,000 Jewish insurgents and committed a series of pointless atrocities. A strong anti-Roman sentiment grew in Judaea and in response Varus was removed from the governorship of Syria.

In AD6, after a period of disgrace, Varus was appointed governor of Germania and was sent to the northern borders of the Empire. His main task was to organise the newly conquered lands into a recognisable Roman province, in order to formalise their annexation to the Empire. Considering his personal background and previous career, Varus was clearly the wrong choice. He had little military experience, no knowledge of Germania and had proved he was a terrible administrator. The governorship of Germania was a second chance for him and for his career, which he was given because of his personal connections. Augustus knew of the risks associated with this appointment and sent Arminius to Germania to act as an expert and reliable adjutant to Varus.

When Varus arrived in Germania, the Roman military forces deployed on the Rhine had been greatly reduced in number. Most of the units stationed on the borders of Gaul, in fact, had been sent to the western Balkans in order to fight against the local insurgents. As a result, only three legions were available for campaigning in Germania (the XVII, XVIII and XIX). The latter did not have a very long military tradition, since it had been formed during the last phase of the Roman civil wars. In addition, with the exception of the XIX, the legions had never fought for long periods in Germania.

Arminius soon became a trusted advisor to Varus, who knew nothing about Germania and had to rely on Arminius to make the most important operational decisions. After a few months of governorship in Germania, Varus illustrated how little he had learned from his previous mistakes. He started to impose heavy taxes on the Germanic tribes and began using harsh repressive methods to ensure the loyalty of his new subjects. The insolence and cruelty of Varus were intolerable for the Germani, who still considered themselves to be free individuals, and who had accepted the Roman

A Roman legionary on the march, wearing a straw sun hat. (Photo by Marc Seriol (@marcmarkhus_photo), Legio II Traiana Fortis – Cohors I Barcinonum, Barcino Oriens)

presence in their territories only for practical reasons. The taxes imposed by the new governor were impossible to sustain and the violence of the occupiers caused the death of many civilians.

In a few months, the Germani had lost their beloved freedom and had found themselves with an army of occupation in their homeland. Arminius was disgusted by the conduct of Varus and by his love for gold. He also realised that the Germani would never flourish under such an inept governor and decided that it was better for them to be independent. Despite having spent most of his life in Rome, the Germanic prince still loved his homeland and its inhabitants: Roman education had not cancelled his 'true' Germanic spirit as Augustus had hoped. The Romans had, in fact, taught all their military practices to a man who was about to become one of their worst enemies. Learning from recent history, Arminius understood that the Germani had never been able to prevail over the Romans because they had always been fragmented politically and militarily. Now, with Varus in Germania, the only way to defeat and expel the Romans was to create a large 'inter-tribal confederation' of Germani that would have been able to deploy a massive Germanic army.

A Roman legionary in full marching outfit, including sun hat. (Legio XI C.P.F. Hispaniensis)

Arminius had secret meetings with the leaders of the major Germanic tribes, and with his charismatic personality was able to convince them to create a large alliance. Most of the Germanic tribes were divided by strong rivalries, and it was not easy for Arminius to override the memories of the recent past in order to propose a new vision to all the Germani. However, six major tribes of western Germania decided to support Arminius's plans: the Cherusci, the Marsi, the Chatti, the Bructeri, the Chauci and the Sicambri. These were supplemented by the surviving elements of the Suebi who still lived in Germania and who had not been crushed by Caesar. After cementing the Germanic inter-tribal alliance, Arminius waited for the right moment to strike while continuing to act as Varus's main adjutant.

In AD9, during the early days of September, Varus moved with his three legions from a newly-built summer camp on the Weser River to his winter headquarters located not far from the Rhine. During the march, Varus was informed of the outbreak of a local revolt that was taking place not far from his actual position; both the uprising and the reports of the rebellion were fabricated by Arminius, who wanted to make the legions leave their usual path and enter a densely forested area where he could ambush them with his Germanic warriors. Varus ordered Arminius to guide him to the area of the rebellion, since he did not have a proper knowledge of the terrain that the legions were crossing. Arminius directed the Romans along a route that was unfamiliar. Another Cheruscan nobleman, a personal enemy of Arminius who was named Segestes, tried to warn Varus that the Roman legions could be ambushed and accused Arminius of being a traitor. Segestes's warnings were dismissed by Varus, who thought them the result of the personal feud between the two Cheruscan noblemen. After these events, Arminius left the Roman column under the pretext of assembling Germanic forces to support the Romans in crushing the ongoing rebellion.

In reality, the Cheruscan leader went to the defensive positions that his Germanic warriors had prepared in the surrounding woods and assumed command of his military forces. The deadly ambush organised by Arminius took place at Kalkriese Hill, in Lower Saxony, not far from present-day Detmond. Varus's troops comprised the three legions mentioned above as well as six infantry cohortes and three cavalry alae of auxiliaries. Like most of the legionaries, they did not have great combat experience. The Roman soldiers did not march in combat formation, since they believed they were in safe territory and would not be attacked. In addition, the various military units were interspersed with hundreds of civilian camp followers. These included the families of the legionaries and of the auxiliaries, as well as those merchants and other workers that lived together with the soldiers in their camps. In total, the Roman military forces comprised around 19,000 soldiers, who marched together with 4,000–5,000 non-combatants.

Following Arminius's instructions, the Roman column started to follow a path that crossed a densely forested area. As it continued to advance, the path became increasingly narrow and muddy. Expecting no attacks from the local Germanic tribes, Varus did not send a cavalry explorer ahead of his main column and continued to march even after a violent storm began. The Roman column was gradually and dangerously becoming too long and too slow due to the broken morphology of the terrain it was crossing. By now it was more or less 20km long, and the units marching at the head of it could not communicate rapidly with those marching at the back. When the legions were stretched and dispersed among the woods, the Germanic warriors emerged from the dense vegetation of the forest and started to attack the Romans. Taken by surprise, the Romans did not have time to redefine their defensive formation; in any case, the limited space available along the narrow path they were following made the adoption of any combat formation practically impossible. Arminius had an extended knowledge of the Roman military tactics and had chosen the perfect location to organise his ambush. He ordered his warriors not to attack the Romans in a direct way, at least during the early phase of the battle. They had, in fact, to skirmish with the legionaries from a distance using light javelins. In this early phase of the clash, hundreds of Romans were killed without even seeing their enemies.

The Germani attacked single portions of the long Roman column and so always had a decisive local superiority over their enemies. Any small group of legionaries who became isolated from the main column was rapidly surrounded by the Germanic warriors and easily massacred. By the end of the first day of battle, the Romans had lost thousands of soldiers without causing significant losses to their enemies. Under heavy rain, Varus ordered the construction of a temporary camp in the forest and spent the night in this newly fortified position. On the following morning, the Romans tried to leave the woods to reach an open area that was located not far from their temporary camp; the legionaries, however, were attacked by Arminius's men and were unable to leave the forest. The torrential rain did not help the Romans, who panicked and had no idea how to deal with their dramatic situation.

At the end of the second day of battle, during the night, the Romans marched towards the foot of Kalkriese Hill in the hope of regrouping on higher ground and putting up stronger resistance against the Germani. Arminius, however, had already foreseen this move and had prepared a strong defensive position at Kalkriese Hill. To climb over it, the Romans had to abandon the main path and follow a sandy strip between the same hill and a nearby swamp that was known as 'Great Bog' because of its extensive dimensions. Between Kalkriese Hill and the Great Bog there was a small gap of only 100 metres, which had dense vegetation on the left (at the foot of the hill). Upon reaching this location, the Romans saw

that the new path was blocked by a trench, making it impossible for them to reach the top of the hill. At the same time, they learned that the Germani had built a solid earthen wall in the dense vegetation that covered the foot of Kalkriese Hill on their left. This earthwork, located on the roadside, permitted the Germani to attack the Romans from an advantageous position with their throwing weapons. Arminius had prepared the most perfect ambush in the military history of the Ancient World: Varus and his men had no chance of escape. Behind the Roman column, 10,000 Germanic warriors had followed and were ready to prevent any retreat. At its front, the narrow path disappeared into the forest where the vegetation was too dense to permit any further advance; on the right, the Romans had the Great Bog; and on the left, there was the earthen wall built by Arminius and defended by 14,000 Germanic warriors.

The Germani attacked the Romans with increasing violence and killed hundreds of them without leaving their fortified position. The legionaries made a desperate attempt to storm the earthen wall built by their enemies, but were repulsed with severe losses. At this point, the few cavalry of the Roman column tried to flee from the trap by abandoning the foot troops, but the mounted auxilia were soon intercepted and massacred by the larger Germanic cavalry that Arminius had kept in reserve. When it became apparent that the Roman formations were collapsing, the Germani came out from their positions and launched a direct attack against the surviving legionaries. The ensuing hand-to-hand fighting was particularly violent and

The rear view of a Roman legionary in full marching asset. (Photo by Marc Seriol (@marcmarkhus_photo), Legio II Traiana Fortis – Cohors I Barcinonum, Barcino Oriens)

Roman legionary (left), officer of the auxiliaries (centre) and *Optio* (right). (David Burns, Legio XIIII Gemina)

dramatic, since the Roman soldiers had no more chances of survival and could only save their honour by putting up an heroic last stand.

Varus, having understood that the disaster had been produced by his ineptitude, committed suicide. His soldiers grouped themselves in improvised defensive formations and tried to resist as long as they could. They were attacked by thousands of Germani, who slaughtered them one by one, together with all the civilians who were part of the Roman column. Most of the Roman officers committed suicide, preferring this to capture and torture. In total, more than 20,000 Romans were killed in the clash that would later be known as the Battle of the Teutoburg Forest. This was one of the worst defeats ever suffered by the Roman Army, which lost three of its formidable legions in an obscure and unknown forest in central Europe. The few surviving Romans were all captured and enslaved by the Germani, except for the officers who were sacrificed during the religious ceremonies that celebrated the great Germanic victory. Very few Romans were ransomed and could return to the western bank of the Rhine, where they reported the terrible events of Teutoburg to the local authorities.

Arminius had obtained a complete victory over the Romans and had practically destroyed all the Imperial military forces that were deployed in Germania. His warriors took from the dead legionaries and auxiliaries thousands of weapons/cuirasses, in addition to the goods and money transported by the many civilians who were part of the Roman column. When news of the Battle of Teutoburg reached the Rhine, the Roman authorities had no choice but to evacuate all the forts and garrisons that had been recently built on the eastern bank of the river; the same territory of Gaul was now exposed to the raids of the Germani. Only one Roman fort, built at Aliso, resisted for several weeks and repulsed all the attacks launched by the Germani, but in the end its garrison was forced to retreat to the Rhine like all the other Roman military units, although it had gained enough time to enable the legions that were stationed in Gaul to organise a 'defensive line' along the river. In that moment, the Romans had just two legions on the frontier between Gaul and Germania. As a result, an invasion of the Gallic territories by Arminius was not impossible. The great Germanic leader, however, was fully satisfied that he had reached his primary objective: freeing the whole territory of Germania from the Roman military presence.

Upon hearing of the terrible defeat suffered in Germania, Augustus was so shocked that he spent several days and nights butting his head against the walls of his palace. According to the Roman historian

Above left: A Roman superior officer wearing parade helmet with mask and muscle cuirass. (Legio XI C.P.F. Hispaniensis)

Above right: Roman centurions wearing lorica segmentata. (Legio XI C.P.F. Hispaniensis)

Suetonius, the Emperor continued shouting the same phrase with a desperate tone: '*Quintili Vare, legiones redde!*', ('Quintilius Varus, give me back my legions!'). Augustus spent the last five years of his life haunted by the massacre caused by Varus, which became commonly known as *Clades Variana* in Rome. The three legions that had been destroyed by Arminius were never reformed, because the Romans were extremely superstitious and did not want to give to a new military unit the official denomination of a corps that had been wiped out in such terrible circumstances.

After the Battle of Teutoburg, Arminius became the supreme leader of western Germania and increased his personal influence over those Germanic tribes that had not been part of his confederation. The Cheruscan leader considered his victory over the Romans only as the starting point of a larger process that would have ended with the political unification of Germania. Arminius knew that the Romans would soon attack the Germani to have their revenge: they would assemble a very large invading force and avoid the mistakes committed by Varus. Attempting to establish a military alliance with the powerful Marcomanni, whose leader Maroboduus was the only other 'supreme' overlord of Germania, Arminius sent the severed head of Varus to them together with the offer of an anti-Roman alliance. Maroboduus, who had recently been recognised as king of his tribe by the Romans, refused Arminius's offer and sent Varus's head to Rome to show his respect for the Empire.

Meanwhile, the Romans sent Tiberius to Gaul with three legions and prepared themselves for a massive counter-offensive. The honour of the Roman Army had been offended and the Imperial Eagles of Varus's legions were in enemy hands. The powerful Marcomanni remained neutral during the ensuing conflict between Arminius and the Romans. As a result, the Cheruscan leader had to fight alone against a superior Imperial military force. From a historical point of view, the Battle of Teutoburg had enormous importance, since it marked the end of the short-lived Roman presence in Germania. Following Arminius's victory, the Rhine began to mark the border between two worlds and two civilisations: the Roman one and the Germanic one.

This cultural division of Europe is still visible today, in the languages spoken by the European peoples and in their way of thinking. No one knows how the history of the world may have been different if present day Germany had been 'romanised'. What we do know is that the Germanic invasions of the Roman Empire would never have taken place and that the fall of Rome would have been completely different from the one we know (which was mostly the result of the great Germanic migrations).

Chapter 3

From Tiberius to Nero, AD14–69

Following the Battle of Teutoburg, Germania was in turmoil: the western part of the region was dominated by Arminius, who was now waiting for the return of the Romans coming to take their revenge; and the eastern part was dominated by the Marcomanni who had transferred their communities to the territories of the present day Czech Republic. Soon after achieving his great victory over the Romans, the Cheruscan leader had tried to form an alliance with the Marcomannic

A Roman centurion with lorica hamata. (Legio VI Victrix Cohors II Cimbria)

king Maroboduus, but the latter had refused. This political act, unexpected by Arminius, led to a progressive increase in tension between the Cherusci and the Marcomanni. Germania could not be ruled by two overlords, but for the moment the inter-tribal wars were postponed in view of the legions' return. Augustus was furious because of the Clades Variana and soon after the events of Teutoburg he had sent Tiberius, together with significant military forces, to Germania. Apparently, the Emperor wanted to attempt a reconquest of the areas that had been lost, but his future successor had other plans for the Germanic campaign that he was about to conduct.

Upon reaching the Rhine in AD10, Tiberius decided to simply stabilise the eastern frontier of Gaul and did not attempt to crush Arminius's forces in their own territory. The Rhine fortifications were repaired and supplemented by new defensive structures; the available garrison forces were distributed in a more effective way and the whole frontier region was militarised. After securing his supply lines, Tiberius moved into Germania but just to conduct some punitive long-range raids: the Romans burned houses, devastated fields and routed all those local Germanic forces that tried to stop them. Arminius, understanding that the Romans had decided to not attempt a new invasion, remained quiet and did not attack the raiders. Tiberius moved very slowly during his new campaign in Germania and took all the necessary precautions: he was terrified of being ambushed like Varus and so did not deploy a very courageous strategy. His advances were extremely cautious, all orders were given in writing, and he had to be consulted for any decision (even the smaller ones). Clearly, Tiberius was obsessed by the possibility that a traitor could cause the defeat of his military forces.

In AD14, Augustus died and Tiberius returned to Rome to become Emperor. During the previous years, his military actions in Germania had achieved very little, except for punishing some local communities that had supported Arminius. Command of the Roman military forces on the Rhine passed to Nero Claudius Drusus, the adoptive son of Tiberius who was nicknamed 'Germanicus' (15BC–AD19) because of the great victories he achieved in Germania. Germanicus was the son of Drusus, one of the first Roman generals to campaign in Germania. Drusus was a younger brother of Tiberius, so Tiberius was the uncle of Germanicus. The second Emperor of Rome had a predilection for his adoptive son Germanicus, to the point of choosing him as his successor instead of his own son Drusus Iulius Caesar (14BC–AD23). Germanicus was much more aggressive than Tiberius from the beginning of his period of command in Germania and soon earned a very good military reputation. He lived together with his soldiers and faced all the same difficulties encountered by his men; for these reasons and for his simple way of life he was particularly loved by the legionaries.

With an impressive force of 12 legions under his command, during the years AD14–16 Germanicus attacked the confederation of Arminius and obtained a series of brilliant victories over it. The new Roman commander did not want to invade Germania, since this was by now considered impossible by all Romans; he wanted, instead, to fight a great pitched battle against the Germani in order to crush them and purge the terrible memories of Teutoburg from the hearts of the Romans.

In addition, Germanicus had another two objectives: finding the three Imperial Eagles that had been captured from Varus's legions and to kill, if possible, Arminius. The Romans had understood that with the death of the Cheruscan leader the partial unity of the Germani would have soon ended. The divided Germanic tribes would have remained impossible to conquer, but they would have been in no position to menace the frontiers of the Empire (at least for the moment).

Germanicus mounted a rapid offensive and attacked the lands of the Chatti and of the Cherusci. He went to the site of the Battle of Teutoburg and buried the bones of the Roman soldiers that still lay there several years after the clash. This was extremely important symbolically, especially for the morale of the Roman Army. During this early phase of his three-year Germanic campaign, Germanicus found one of the three Imperial eagles. The Germani did not try to stop the movements of the legions and simply conducted minor skirmishes to harass their enemy. With the progression of time, however, Arminius was forced to change his strategy: the Roman legions were destroying and pillaging everything they encountered on their march and this was damaging his personal reputation as overlord of the Germanic confederation. Many of the tribes who had followed him with enthusiasm were now ready to rebel against his rule and so he had no choice but to fight Germanicus in a field battle.

The Battle of Idistaviso

The battle took place at Idistaviso, on the right bank of the Weser River, between present day Minden and Hamelin. Arminius deployed his forces in an intelligent way, since between him and the legions there was the Weser River: he was on the right bank, Germanicus was on the left bank. Crossing the river under a rain of missile weapons thrown by the Germani would have been extremely difficult for the legions. Germanicus, however, could count on large contingents of auxiliaries coming from Gaul and from communities of Germanic allies (the Batavi and the Chauci) who had joined the Roman Army. Covered by the cavalry that advanced in two columns, the bulk of the Roman forces was able to cross the Weser without experiencing major difficulties. One of the two Roman cavalry columns, mostly made up of Batavi, advanced deeply into enemy positions so that they lost contact with the rest of Germanicus's troops. The column was surrounded by the Cherusci and risked being completely destroyed. At a certain point, however, it linked up with the rest of the Roman cavalry and was able to retreat.

The main Germanic army, commanded by Arminius, left its initial positions after being unable to stop the Roman crossing of the Weser. During the following night, the Romans built a large fortified camp in the centre of the battlefield; the Germani attacked it, hoping to be helped by darkness, but all their assaults were repelled. On the following morning, the decisive clash between the two armies took place: the Germani attacked first but were repulsed with very high casualties. At this point of the battle, the two Roman wings, made up of cavalry and auxiliaries, encircled the enemy and attacked it on the flanks. Arminius, seriously wounded, had no choice but to flee from the battlefield with the

Above left: A Roman centurion with lorica hamata. (Marc Seriol (@marcmarkhus_photo), Legio II Traiana Fortis – Cohors I Barcinonum, Barcino Oriens)

Above right: A Roman centurion with hexagonal shield. (Legio XIII Gemina)

few survivors of his army in order to escape capture. The Battle of Idistaviso had been Germanicus's masterpiece and had shown to the world that the Roman Army was still invincible on the open field. Arminius was no match for it.

Following their victory, the Romans occupied a strong defensive position that had been built by the local Germanic tribes not far from the battlefield. Known as the Angrivarian Wall, this marked the border between the territories of the Angrivarii and those of the Cherusci. Expecting a Germanic counter-attack, Germanicus occupied it and waited for the arrival of the enemy. The Germani attacked the earthen wall shortly after the Romans occupied it, but all their assaults were repulsed with severe losses. After this second defeat, Arminius could no longer counter the actions of the Romans in Germania and had to deal with the rapid crumbling of the inter-tribal confederation that he had forged. The lands of the Chatti and of the Marsi were devastated by Germanicus, who managed to find another of the three Imperial eagles lost at Teutoburg. After having reached all their objectives except for the capture of Arminius, the Romans returned to the Rhine and left Germania without attempting an occupation of the territory. Their revenge had been cruel and had destroyed any possibility of Germanic political unity.

In AD17, hoping to take advantage of the weakness, the Marcomanni of Maroboduus attacked the Cherusci. This inter-tribal war, probably inspired by the Romans, ended in a stalemate since the Marcomanni was not able to achieve any significant victory. A few years later, both the main protagonists of the Battle of Idistaviso died: Germanicus was poisoned by his opponents during AD19, while he was in Antioch. Arminius was assassinated two years later by a rival nobleman of his own Cheruscan tribe. With the death of Arminius, any hope of Germanic political unity vanished. The great warrior leader, however, had prevented the possibility that the Romans would ever conquer Germania. In AD18, meanwhile, Maroboduus was replaced as king of the Marcomanni by a younger noble who was supported by the Romans. He would die in AD37, as an exile, in Italy.

Pannonia

In the last years of Augustus's reign and the following ones of Tiberius's early reign, the Romans consolidated their presence in Pannonia. Pannonia had great strategic importance, since it was located north of the western Balkans, where important commercial routes connected the western territories of the Empire with the eastern ones. After the death of Julius Caesar, hoping to gain some advantages from the chaotic political situation in Rome, the Pannonians had supported an Illyrian military revolt and had defeated – together with their allies – the Roman military garrisons stationed in Illyricum. To restore order in the area, during 35–33BC, a young Augustus (not yet Emperor) campaigned against the Pannonians and the Illyrians. During this conflict, the first Roman incursions into Pannonia were conducted and the southern portion of the country came under a strong Roman influence (albeit not being occupied in a permanent way).

During 17–16BC, the Celts of northern Pannonia allied themselves with the Kingdom of Noricum, located on the western border of their territories and resumed hostilities against Rome. The Kingdom of Noricum was originally inhabited by several Celtic tribes, which were gradually unified into a sort of 'federal state' that was strongly influenced by Rome. The Celts of Noricum had strong political and commercial relations with those living in Pannonia (contemporary Hungary) and with the other Celtic tribes settled in Switzerland. Noricum was in a delicate geographical position, with Roman Italy to the south and the fierce Germanic nations to the north. As a result, its political leaders had no choice but to form an alliance with the Roman Republic in order to obtain the military protection of that nation. The Kingdom of Noricum was created around 150BC, and since its inception it provided the Romans with large amounts of excellent weapons. In fact, Celtic Austria was famous for the production of high quality metal weapons and tools. Noricum was rich in iron, gold and salt. Both the Romans and the Germani required these natural resources, so the local Celts had to maintain some kind of 'equilibrium' in order to preserve their independence.

A Roman centurion wearing the distinctive helmet with transverse crest and muscle cuirass. (Legio XI C.P.F. Hispaniensis)

Noric steel was particularly appreciated for its quality and hardness: most of the weapons used to equip the Roman legions were obtained from it. During the early years of Tiberius's reign, the Austrian Celts changed their attitude towards Rome and joined their Pannonian cousins in an invasion of Roman Macedonia. It was a complete failure: the invaders were promptly defeated and – in retaliation – the Roman legions invaded the Kingdom of Noricum (annexing it to the Empire).

The northern part of Pannonia remained independent after these campaigns. In 14BC, the Illyrians of southern Pannonia rebelled against Roman rule, with the help of the Celts living in the northern part of the country. The new war lasted until 10BC and ended with the Roman reconquest of southern Pannonia. In addition, some northern Celtic territories were occupied by the legions. It was clear by now that Roman dominance in the region was becoming absolute; the only chance for the Celts and the Illyrians to survive was to unite against the common menace to regain some form of independence. Their last hopes for freedom were crushed during the Great Illyrian Revolt. Pannonia became a Roman province in AD14, just a few years after order was completely restored in both Illyricum and Dalmatia.

Thrace

During the first decades of the Imperial period, the Roman Army was involved in several minor military operations that were fought in Thrace and in Dacia. The eastern Balkans were still mostly free from Rome's political and military influence. In 150BC, an attempt to re-establish an independent Kingdom of Macedonia provoked the outbreak of the Fourth Macedonian War, which lasted until 148BC. During this conflict, as in several previous ones, the Thracians were part of the anti-Roman front. The pretender to the Macedonian throne, however, was defeated by the Roman Republic and in 146BC Macedonia was officially transformed into a Roman province. Since that year, a state of constant war existed on the border between Roman Macedonia and independent Thrace. The Romans wanted to transform the various Thracian tribes into client communities since they considered themselves as the heirs of Macedonia's political influence over Thrace. The Thracians, of course, had no intention of accepting the Romans as their overlords and soon started to launch aggressive raids against the province of Macedonia.

During the many little wars fought against the Thracians, the Roman Army was defeated on several occasions and one of its proconsuls was even killed. The Thracian warriors, with their elusive skirmishing tactics and great mobility, proved to be very difficult to fight and the Romans' lack of knowledge of the eastern Balkan territory put them at a significant disadvantage. From 146BC, the Roman Republic had tried to re-create a unified Thracian state, in order to transform it into a vassal kingdom. It would have been much easier for the Romans to control a single 'puppet kingdom' instead of keeping order among many war-like tribes.

Around 100BC, a new Thracian tribe was organised into the Odrysian Kingdom, but this did not last for long due to the internal divisions of the Thracians (most of whom were strongly against Rome's indirect rule of their homeland). Around 30BC, some form of Thracian Kingdom had been restored by the Romans. By this date, the Romans and the still autonomous Thracian tribes had all accepted some form of Roman suzerainty. In 15BC, the population of the Thracian Kingdom rose in revolt against the Romans and killed the puppet monarch who had been chosen by the foreigners. Very soon, however, the rebellion was crushed by the legions. New Thracian uprisings took place during the following decades, as the Roman presence in what is now present day Bulgaria became increasingly stable.

In AD12, the Romans divided the territories of Thrace into two puppet kingdoms, but this measure did not change the situation. The two Thracian realms soon started to fight each other and anti-Roman uprisings occurred with great frequency. In AD45, a new large rebellion erupted in Thrace, which caused the death of one of the puppet kings. To stop the endemic warring ravaging the region, Emperor Claudius (10BC–AD54) (fourth ruler of Rome, after Tiberius and Caligula (AD12–41)) decided to transform Thrace into a Roman province (AD46). Thracian uprisings and revolts continued for some years, but without achieving significant results. Like many other peoples, the Thracians had lost their independence forever. During the following centuries, they formed a fundamental component of the Roman Empire's population.

A Roman centurion with scale armour and rectangular shield. (Legio XI C.P.F. Hispaniensis)

Dacia

During the second half of the third century BC, as a result of Celtic migrations across Europe, the Celtic presence in Dacia (modern Romania) became quite significant. The newcomers had to co-exist with the local population just as the nomadic Scythians had already done before them in 5BC. In 150BC, the Dacians started to fight the Celtic communities living on their territories with the objective of expelling them from Transylvania. This was not a simple process, because the Celts had established themselves very firmly. The Dacians, however, were supported in their long struggle by the powerful Thracian tribe of the Getae (who controlled a portion of territory located north of the Danube). The violent conflict between Dacians and Celts came to a critical point around 60BC, when a great war leader emerged from the tribes of Dacia – Burebista. Burebista guided the Dacians in the decisive moment of the conflict against the Celts and expelled their enemy from the middle Danube region. Soon after achieving his objectives, Burebista was crowned overlord of the Dacians and started to unify all the tribes of his people into a centralised state. The state developed quite rapidly. Burebista ordered the construction of a system of hill forts across Dacia from which to control the territories of the various communities, and gradually absorbed the powerful Getae into his political sphere of influence.

Until 40BC, he continued to fight on the borders of his new state against the Celtic Boii, a Celtic tribe living in Pannonia, who still represented a potential menace for Dacia and who still launched raids against Burebista's lands. Burebista, after having secured his control over most of the eastern Balkans, started to have an aggressive attitude towards Roman Macedonia and launched several raids against the Illyrian/Thracian tribes living on his southern borders. From 55BC, the Dacians also started to attack the Greek colonies that were located on the Black Sea coastline. The Romans did very little to counter Dacian expansionism during this phase, because they were heavily involved on other fronts. Julius Caesar was completing the conquest of Gaul and the possibility of a civil war between him and Pompey was already becoming a certainty.

Civil war erupted in 49BC and affected the whole Mediterranean, since each state bordering the Roman Republic was forced to take sides in favour of Caesar or Pompey. Burebista supported Pompey, who was defeated and killed before he could send any military contingent into Roman Macedonia. Caesar was well aware of the Dacians' military potential and considered Burebista to be a great enemy. After becoming the sole ruler of Rome, he started planning a large punitive expedition against Dacia but was assassinated (44BC) before the campaign could become reality. Some months after Julius Caesar's death, Burebista was killed during a plot organised by the same Dacian aristocracy, which resented the new centralised form of government introduced by Burebista and which feared the possibility of an armed conflict with Rome. Soon after these events, the Dacian kingdom collapsed and was broken up into several smaller realms that soon started to quarrel between themselves.

Following Burebista's death, the Dacians no longer represented a menace for the Romans. Their large kingdom was now fragmented into several parts and the great multi-national army created by Burebista (numbering around 100,000 warriors) no longer existed. The Roman Republic abandoned Caesar's plans for an invasion of Dacia and limited itself to keeping order on the Balkan frontier. This situation started to change only in AD69 when a new Dacian leader, named Duras (ruled 69–87), emerged. Until the end of Augustus's reign (AD14), Dacia continued to be divided into five smaller kingdoms that could all be controlled easily by the Romans.

In the following decades, however, a new sense of national identity started to emerge among the Dacian communities. Rome was increasingly perceived as a deadly menace and all Dacians began feeling the need for a joint action against the new enemy. Duras became king in AD69, after his father Scorilo (d.70) was probably killed during a first Dacian raid launched against the Roman province of Moesia. The latter had been created by the Romans during Augustus's reign, as a buffer zone between Macedonia and Dacia. Moesia, in fact, was located to the north of Macedonia and to the southwest of Dacia. Before the Romans' arrival, this region was inhabited by the Moesi (a tribe of Dacian stock), the Triballi (a Thracian tribe) and by a portion of the Germanic Bastarnae. The Romans considered Moesia as a key region and decided to conquer it in order to protect Macedonia from the incursions of barbarians living north of the Danube. The northern border of the new province, in fact, was to be marked by the course of the great river. The *casus belli* for the invasion of Moesia was provided to the Romans by the Bastarnae, who attacked one of their allied Thracian tribes. During the ensuing conflict, thousands of Bastarnae were killed by the Romans and the Moesi were forced to submit. Before reinforcements sent by the northern Bastarnae could come south of the Danube, the Romans had already completed the conquest of Moesia, which was later transformed into one of their provinces (AD6). The Dacians never accepted Rome's conquest of Moesia and soon started to attack Moesia with rapid incursions. The first major raid was organised by Scorilo in AD69 but was a failure. For the moment, however, the Romans had no intention of fighting a large-scale war against the Dacians and continued to adopt a defensive strategy on the Danubian limes.

Above left: A Roman centurion with scale armour. (Legio VI Victrix Cohors II Cimbria)

Above right: A Roman centurion with scale armour and greaves. (Legio XI C.P.F. Hispaniensis)

Anatolia

During the republican period, Anatolia (modern Turkey, known as Asia Minor by the Romans) saw the presence of three small political entities that co-existed with the Roman Province of Asia and included the kingdom of Bithynia, the kingdom of Cappadocia, and Galatia. The Bithynians were one of the Thracian tribes that had migrated from Europe to Anatolia, as the more numerous Phrygians before the fifth century BC. Despite being formally subjects of the Persian Empire, since 435BC (well before the arrival of Alexander the Great [356–323BC] in 334BC) they had started to act as a sort of independent realm. Since 297BC, the rulers of Bithynia started to call themselves kings, becoming fully autonomous in 281BC. Nicomedes I (300–255BC), ascending to the Bithynian throne in 280BC, was the first monarch to launch a significant process of 'hellenisation' in his realm: until then, in fact, the autonomous Bithynians had continued to fight as light infantrymen like every Thracian tribe. In 74BC, the last monarch of Bithynia, having no heirs, bequeathed his kingdom to Rome.

A Roman cavalryman with parade helmet and greaves (left) and optio with crested helmet and scale armour (right). (Legio XIIII Gemina)

The independent kingdom of Cappadocia was founded in 331BC by Ariarathes (405–322BC), the last Persian governor of the Cappadocian Satrapy. Since the Macedonians continued their advance towards the heart of the Persian Empire without attacking Cappadocia, Ariarathes was able to retain power in his own province and avoided a direct confrontation with the army of Alexander the Great. Ariarathes was defeated and killed by one of Alexander's successors in 322BC, but after a brief period of Macedonian occupation, Cappadocia became independent again in 301BC under Ariarathes's son. The latter initiated a new royal family that was to last until 96BC. In that year, in fact, Mithridates VI (135–63BC) of Pontus invaded Cappadocia and briefly annexed it to his possessions. After having defeated Mithridates, however, the Romans decided to install a new royal family in Cappadocia and transformed the small realm into a client kingdom. In 36BC, the last exponent of this new dynasty died without heirs and thus the Romans gave Cappadocia to a local noble named Archelaus (23BC–AD18), who was a personal friend of Mark Antony. When Archelaus died in AD17, Cappadocia was absorbed into the Roman Empire.

The Celts

When the Celts invaded the southern Balkans in 280BC a number crossed the Hellespont and went to Anatolia. Apparently, the Celtic invaders went to Asia at the invitation of Nicomedes I, King of Bithynia, who wanted their military help in a dynastic struggle with his brother. The Celts, settling in the centre of Anatolia, belonged to three different tribes: the Tectosages, the Trocmi and the Tolistobogii. These settled in the plateau of Phrygia, after submitting the local inhabitants of Thracian descent. Very soon after, this region (now Ankara) started to be known as Galatia. The three tribes organised themselves into a loose tribal federation, which exerted control over the subject Phrygian peasants. During the early phase of their settlement, the Galatians mostly supported themselves by plundering bordering countries or by serving as mercenaries in the various Hellenistic armies of the time.

In 232BC, the Attalids of Pergamon defeated them in battle: this eventually led to the creation of a more permanent Galatian settlement in central Anatolia and became a vassal of the Hellenistic kingdom of Pergamon. In 189BC, after transforming the kingdom of Pergamon into their province of Asia, the Romans launched a large expedition against the Celts of Anatolia, which became known as the Galatian War. After being defeated, the Galatians lost much of their military power and were later invaded by Mithridates VI. Thanks to the decisive help of the Romans, however, the Celts of Anatolia were later able to regain their independence after the end of the Mithridatic Wars (88–63BC) that were fought between Pontus and the Roman Republic.

In 62BC, Galatia formally became a client state of Rome and was officially organised as a kingdom. In 25BC, this kingdom of Galatia was annexed by the Romans. Since 62BC, Galatia had deployed a regular army organised according to contemporary Roman models and trained by Roman military officers. Deiotarus (120–41BC), who was made king of Galatia in 62BC as a reward for the military support given to Pompey, re-structured the military forces of his realm as 30 cohortes (the equivalent of three Roman legions) and levied a total of 14,000 soldiers (12,000 infantrymen and 2,000 cavalrymen). These took part in the wars against Mithridathes VI as loyal allies of Rome, but after some severe defeats had to be reduced to a single legion. This remaining legion, however, continued to fight on the side of the Romans during Caesar's brief campaign against Pontus of 47BC. After the death of Caesar, the Galatians made a crucial mistake: they sided with Mark Anthony and lost their independence after he was defeated in 30BC. With Galatia's incorporation into the Roman Empire, the single legion of that kingdom was absorbed into the Roman Army as the 'Legio XXII *Deiotariana*' (from the name of the king who had founded it).

The Roman invasion of Britain

In Britain, the Iron Age began around 600BC, one century before that in Ireland. It was during this period that the Celtic influence started to strengthen in the British Isles. The Celticisation of the Isles was not characterised by mass migrations or by violent campaigns of conquest: certainly a good number of Celts arrived in Britain from northern Gaul, but these settled on the British territory in a passive way and soon mixed with local populations, who were already organised in tribes. Over time, especially during the fourth century BC, the whole population of Britain assumed the main characteristics of the Celtic 'La Tène Culture'. The presence of the early immigrants coming from Gaul had been fundamental in this sense, since a quite limited number of settlers had been able to influence and change the civilisation of the British Isles. The process was slower in Ireland, Scotland and Wales but gave more permanent results than in Britain. By the end of the fourth century BC, the whole of the British Isles had become the home of hundreds of Celtic tribes: only the territory of Gaul could be considered as 'more Celtic' than that of Britain.

After his early victories over the Celts of Gaul, Caesar turned against the Britons in order to punish them for the support given to the Celts of Armorica (Brittany) during the previous years. The Romans disembarked in southern England with two legions, but this time Caesar's campaign was a half-failure: bad weather destroyed a large part of the Roman fleet and the landings were opposed very effectively by the Britons, who employed war chariots on a large scale, something that the Romans were not prepared to face.

In 54BC, Caesar returned to Britain at the head of a larger military force and this time made all the necessary preparations to face the Britons on equal terms. The Catuvellauni, the most important Celtic tribe living in southern England, was defeated and obliged to pay a yearly tribute to Rome. Caesar's expeditions in Britain had not given new territories to Rome, but were extremely important for the propaganda of the famous general: no other Roman military commander, in fact, had ever moved so far north, to the point of reaching what were the 'edges of the known world'.

Above left: A Roman centurion with lorica segmentata and rectangular shield. (Legio XI C.P.F. Hispaniensis)

Above right: A Roman centurion mounted on his horse. (Legio XI C.P.F. Hispaniensis)

Around AD40, only the Celts of Britain and Ireland remained independent from Rome. Their relative geographical isolation meant they were able to preserve their identity from the great political and cultural changes taking place in Europe at that time. Once the Roman Empire was consolidated in Continental Europe, the Romans could start planning an invasion of Britain. Following Caesar's expeditions, the Celts of southern Britain had continued to send tributes and hostages to Rome in order to retain their independence. In the times of Augustus, however, it became clear that the Romans would have occupied the island at the first opportunity. In AD43, Claudius, the third Emperor of Rome, finally ordered the invasion of Britain. It was conducted by an army of 40,000 men, half of which were professional soldiers from four different legions (II Augusta, IX Hispana, XIV Gemina and XX). From a political point of view, the Roman invasion of Britain was determined by Claudius's will to obtain control over the important natural resources of the island noted by Caesar. From a military point of view, it was determined by the need to protect northern Gaul from possible Celtic incursions coming from the British Isles.

The early resistance of the Britons was guided by the Catuvellauni, the powerful confederation of tribes that had already fought against Julius Caesar. The decisive battle of this first phase of the campaign was fought at the River Medway, where the Celts assembled their forces to stop the Romans who had landed at Richborough, on the east coast of Kent. Upon hearing of the Roman landing, the Britons of southeastern England assembled a very large military force – probably numbering around 100,000 warriors – and put it at the orders of two warlords of the Catuvellauni tribe: the two brothers Togodumnus (d.AD43) and Caratacus (AD15–54). The objective of the Britons was one: stopping the Roman advance by using the River Medwey as a natural barrier. For the invaders crossing the river in front of a large enemy force and under a rain of arrows/javelins would have been tactical suicide. No bridge crossed the River Medwey and thus the Britons were sure of their final victory. The Romans, however, could count on a very special military resource ignored by the Celtic leaders. The invaders included a large contingent of Germanic auxiliary cavalrymen, recruited from the Batavi tribe, who were specifically trained to cross rivers wearing their heavy chainmail. The Batavian cavalrymen crossed the Medwey very rapidly and launched a surprise attack against the Britons, engaging their war chariots.

Taking advantage of the enemy's surprise, the *Legio II Augusta* also crossed the river under command of Vespasian (AD9–79) (future Emperor). The Britons were not able to mount an effective counter-attack, and by the end of the first day of battle, the whole Roman expeditionary force had crossed the Medwey. During the following days, the Romans resumed their attacks, inflicting severe losses on the Britons, who resisted with great determination, but ultimately had to abandon their positions. Those living in Britannia were lightly equipped and had never fought against a massive heavy infantry force consisting of four legions. As a result, their defeat on the Medwey was almost inevitable. After this first major battle, the Britons fell back to the Thames, where they soon organised a new line of defence. The Roman troops, however, moved very rapidly and attacked the Britons, some of whose warriors were still crossing the river. The invaders were able to build a temporary bridge across the Thames and so the new Celtic defensive line crumbled without seeing a major pitched battle. During the minor skirmishes that took place in the Thames area, the warlord Togodumnus died. This was a major blow for the Celtic communities, which were now in a very chaotic situation politically. Several tribes considered the previous defeats as decisive and thus wanted to make peace with the Romans. Others wanted to continue fighting using guerrilla methods in order to halt the advance of the invaders. At this point of the campaign, Emperor Claudius went to Britannia, having understood that the latter was on the verge of being conquered. A total of 11 tribes surrendered to Claudius without putting up further resistance; the Romans had occupied most of south-eastern England, but to the north and the west there were still several war-like communities with intact military capabilities.

The invaders established their capital at Camulodunum (Colchester), but Caratacus was not captured and thus could continue his resistance in the west even after Claudius had left Britannia to celebrate his triumph in Rome. Vespasian, the most brilliant Roman commander serving in Britain, took part in the

A Roman centurion with winter marching dress. (Legio XI C.P.F. Hispaniensis)

A Roman centurion with summer marching dress. He is carrying the vitis – the peculiar vine staff of centurions. (Legio XI C.P.F. Hispaniensis)

invading military forces and advanced westwards: he subdued many tribes and conquered several enemy *oppida* (urban settlements) advancing as far as Exeter. The *Legio IX Hispana*, instead, was sent north towards Lincoln with orders to subdue the local Celtic communities. By AD47, after just one major battle and a few years of campaigning, the whole of present day England had been conquered by Rome and had been transformed into a province. During that year, the new Roman governor of Britannia, Publius Ostorius Scapula (AD15–52), launched a first military campaign against the Celtic tribes of Wales. The latter defended their home territories fiercely and caused several problems for the Romans.

In AD50, the Roman troops finally intercepted Caratacus and his resistance forces, obliging them to fight a pitched battle. Known as the Battle of Caer Caradoc, it ended with the defeat of the two tribes that supported the rebel leader: the Ordovices and the Silures. Caratacus's family was captured, but the warlord escaped again, fleeing north, where the resistance of the Britons continued. Here, however, he was captured by the Brigantes, who were temporarily allies of Rome and who handed him over in chains to the invaders. After these events, Ostorius Scapula died and was replaced as governor of Britannia by Aulus Didius Gallus, who – by using quite harsh methods – was able to subdue the warlike Celtic communities living on the border between England and Wales. The new governor, however, made little progress in extending the Roman presence westwards and northwards. In AD54, following the death of Claudius, Nero became the new Emperor. He was very ambitious and searching for military glory and decided to enlarge the Roman possessions in Britain. Quintus Veranius was appointed as its new governor. He was an experienced administrator and able to organize a successful campaign in Wales, which ended with the subjugation of most of the local tribes.

Quintus Veranius understood that the resistance of the Britons, especially in Wales, was guided by the druids and so attempted to kill as many as possible. In Celtic Britain, the druids were religious guides and political leaders who could assemble substantial numbers of warriors in case of war. In AD60, the Romans destroyed a very important druidical centre located on the island of Mona (modern Anglesey), which deprived the rebels of their leadership. Despite this, however, the final occupation of Wales could not be achieved due to the outbreak of a major rebellion in what is now present day England. Commonly known as Boudica's Revolt from the name of the queen of the Iceni who guided it.

The Iceni were one of the most important Celtic tribes living in south-eastern Britain. They had already fought against Rome in AD47 but had been soundly defeated. Despite this, the Romans permitted them to retain some degree of autonomy. With the ascendancy of Boudica, the Iceni took advantage of the Romans' heavy military involvement in Wales to launch a general rebellion. Boudica (AD30–61) was the female equivalent of Vercingetorix, the leader of the Gauls who had fought against Julius Caesar, a charismatic character able to unite most of the Celtic tribes of Britannia against the Romans. The revolt clearly showed that the Roman presence in Britain was not yet consolidated, since the druids were still the real masters of the territory. Most of the tribes had formally submitted only to gain enough time to reorganise their forces. Initially Boudica's rebel army was victorious on various occasions.

Above left: Roman optio wearing lorica segmentata. He is carrying the hastile or wooden staff that was peculiar to his rank. (Marc Seriol (@marcmarkhus_photo), Legio II Traiana Fortis – Cohors I Barcinonum, Barcino Oriens)

Above right: A Roman optio with lorica segmentata armour. (Legio VI Victrix Cohors II Cimbria)

The Britons' first target was Camulodunum (modern Colchester), which was inhabited by Roman veterans who had decided to settle in Britain after years of fighting. The foreign colonists had mistreated the local Celts in many different ways, for example, by obliging them to pay the expenses for the construction of a temple that was built in honour of the former Emperor Claudius. The Roman veterans, after having understood that the locals were on the verge of revolt, asked for help from the new governor of Britannia (Catus Decianus). The reinforcements sent, however, consisted of just 200 auxiliaries and were too few to stop the Britons. Camulodunum was destroyed after some bitter fighting and the last veterans resisted inside Claudius's temple for two days.

After these events, the *Legio IX Hispana* marched against Boudica, but was routed on the open field and was practically destroyed. The Britons took advantage of this victory to march on Londinium (present day London), a new Roman settlement that was growing quite rapidly as a thriving commercial centre. Boudica destroyed Londinium as she had already done with Camulodunum. The Britons seemed invincible, and the Roman presence in Britannia seemed to be on the point of vanishing. The Romans, however, were able to mount a counter-offensive with all the forces at their disposal. A task force consisting of the *Legio XIV Gemina*, some detachments of the *Legio XX Valeria Victrix* and a large number of auxiliaries marched against Boudica. The decisive clash of the war, known as the Battle of Watling Street (c.AD60/61), took place between 10,000 Romans and around 100,000 Britons. Initially, the Romans did not move from their positions, throwing their heavy javelins against the assaulting enemies and inflicting serious losses; then they dashed forward in a wedge-like offensive formation. The auxiliary cavalry, deployed on the Roman wings, charged against the Britons; who were unable to stop the Roman attack and soon took flight. The escape of Boudica's warriors from the battlefield was made difficult by the fact that the Britons had deployed their wagons on the back of their formation; the Romans gave no quarter to the defeated enemies and killed every man, woman or child they encountered in the Britons' rearguard. A total of 80,000 Celts – warriors and civilians – were massacred in what was the bloodiest battle of AD61. Boudica, defeated in a decisive way, ended her life by poison when it became apparent that her revolt had been crushed.

The years AD61–70 saw the Romans completing their conquest of Wales and expanding their territories towards northern Britain. In AD78, the great military commander Gnaeus Julius Agricola (AD40–93) arrived in Britain as new Roman governor of the province. An experienced leader, his main objective was expanding Rome's territories towards Caledonia (Scotland). Firstly, he defeated the last independent tribes of northern England, including the war-like Brigantes, then pushed his military forces to the Firth of Tay, without meeting serious resistance. The advancing Romans built several forts in the newly conquered areas in order to secure their presence in northern England.

The Celts in Scotland

Ancient Scotland was inhabited by some of the wildest Celtic tribes ever encountered by Rome: these had experienced very little contact with other civilisations in the past and were much more difficult to conquer than the tribes of southern Britain. Thanks to an intelligent use of amphibious tactics, Agricola was able to move along Scotland's eastern and northern coasts. He reached the heart of the eastern Highlands, where he finally met a large Caledonian army in the famous Battle of Mons Graupius. The location of the latter is still unknown, but we know for sure that it was fought in northern Scotland.

The Caledonian tribes had assembled a large army with 30,000 warriors, commanded by the warlord Calgacus. Agricola commanded around 18,000 men (legionaries and auxiliaries). The Celts deployed their military forces on the high ground of Mons Graupius, up the slope of the hill, in a horseshoe formation. On the plain, in front of the Romans, there were numerous war chariots. After a brief exchange of missile weapons, Agricola ordered his auxiliaries (most of whom were Germani) to launch a frontal attack against the enemy. The Caledonians were easily cut down by the Romans and trampled on the lower slopes of Mons Graupius; outflanked by Agricola's cavalry, they were completely routed. The Romans pursued them relentlessly, killing around 10,000 of the enemy.

Before being recalled to Rome in AD84, Agricola built a network of military roads and forts to secure Roman control over both the Highlands and the Lowlands. After Agricola's departure, however, the incoming Roman governors of Britain did little to consolidate the conquest of Caledonia or to protect the communication/supply lines connecting southeastern Scotland with northeastern England. The fortifications were abandoned, and most of the local Celtic tribes rapidly regained their previous autonomy. Gradually, the Romans retreated to the border located between northern England and southern Scotland, where a stable limes (frontier) was established from AD122 following the construction of the famous Hadrian's Wall.

During the following two decades, the Romans attempted to re-occupy at least the Scottish Lowlands, but encountered strong resistance from the Caledonians. After understanding that it would have been impossible to move further north against the Scottish Highlands, in AD142, they established new limes by building the Antonine Wall in order to separate the Lowlands from the Highlands. Despite these efforts, in AD162, the Romans were forced to abandon the Lowlands and retreat to Hadrian's Wall. During the following decades the Romans made at least four significant attempts to reconquer southern Caledonia, but all were repulsed.

Above left: A Roman optio wearing lorica segmentata. (Legio VI Victrix Cohors II Cimbria)

Above right: A Roman optio with lorica segmentata. (Legio XI C.P.F. Hispaniensis)

Chapter 4

From the Flavian Dynasty to Marcus Aurelius, AD69–166

Until AD4, most of present day Israel was ruled as a client kingdom by Herod the Great, who was a loyal ally of Mark Anthony but later supported Augustus when the latter gained the upper hand in the final phase of the Roman civil wars. Herod was a tyrant who killed every individual who might claim his throne. However, he protected the integrity of his state from foreign aggression and retained a degree of autonomy from Rome. After his death, the kingdom was divided into four parts between his three surviving sons and his sister Salome I. This so-called Herodian Tetrarchy did not last long, since the Romans used their military presence in Syria to annexe – one by one – the four small Jewish states to their expanding Empire.

The Great Jewish Revolt

By AD44, all the states of the Herodian Tetrarchy had disappeared and the territory of Judaea could be organised as a Roman territory (autonomous from the nearby Province of Syria). Initially, the Romans respected the laws and the customs of the Jewish people, for example by allowing their new subjects to rest on the Sabbath or by granting them exemption from pagan rituals. During the following decades, however, the situation started to change as the Romans tried to 'hellenise' the Jews by introducing Greek language and pagan culture on a large scale. In AD64, Gessius Florus became procurator of Judaea, an appointment that accelerated the ongoing escalation towards the outbreak of a major rebellion. Florus, in fact, stole riches from the treasure of the sacred Temple of Jerusalem and executed many Jewish traditionalists who tried to defend their own culture from Roman influence. What became later known as the Great Jewish Revolt began as a clash between opposing Jewish factions, since there were confrontations between those Jews who wanted to rebel against the Romans and those who favoured the foreign occupation. Florus reacted to the unrest by sending his soldiers into Jerusalem and by arresting some of the rebel leaders. The Jews, however, were outraged by these actions and took up arms against the Roman military garrisons stationed on their territory. In a few days, the legionaries were expelled from Jerusalem as well as from the fortress of Masada, the strongest in Judaea that had been built by Herod the Great.

The Roman military garrison of Judaea was not particularly strong, since it could count on the support of the legions that were stationed in nearby Syria, if needed. The governor of the latter province, Cestius Gallus (d.AD67), soon assembled an army with 30,000 men (including *Legio XII Fulminata* plus detachments from other legions and many auxiliaries) that was sent to reconquer Judaea. Initially, the Romans obtained a series of victories and killed thousands of rebels; they even reconquered most of Jerusalem, but were unable to occupy its citadel, which was strongly fortified. At this point, the Roman commanders, being surrounded by hostile contingents of Jewish fighters, decided to leave Jerusalem and move towards the sea to wait for reinforcements. During their march, however, the Romans were ambushed at the pass of Beth Horon by 25,000 Jewish insurgents.

The Roman troops now consisted of just 6,000 men and were completely outnumbered; the Jews attacked the Romans from a distance, with stones and arrows. By the end of the clash, the Roman legionaries had been massacred: they had not been able to get into formation due to the nature of the terrain and had lost their usual cohesion. After the Battle of Beth Horon, the rebels equipped themselves with weapons plundered from their dead enemies and were able to enlarge their military forces too. The shocking defeat convinced Emperor Nero to appoint Vespasian as commander of a large expeditionary corps that was assembled to reconquer Judaea; the latter comprised *Legio X Fretensis* and *Legio V Macedonica*, which were later joined by the *Legio XV Apollinaris* from Egypt (commanded by Vespasian's son and future Emperor Titus). Meanwhile, the Jewish leaders formed a provisional government in Jerusalem and began reinforcing the capital's defences in expectation of a massive Roman counter-offensive. Having around 60,000 soldiers under his command, Vespasian used political and religious divisions existing among Jews to his advantage: during AD67 he reconquered most of Galilee, in northern Israel, without meeting serious resistance.

Vespasian made the important coastal city of Caesarea Marittima his main base and methodically proceeded to clear insurgent groups from the Judaean coastline. During this phase of the war, more than 100,000 Jews were killed or sold into slavery. In time, the Jewish defenders of Jerusalem started to experience internal conflicts: the radical faction of the Zealots, in fact, was strongly determined to defend the city until the last man, while the other groups wanted to avoid the destruction of their holy capital. In the end, after thousands of moderate Jews were killed, the Zealots gained full control over Jerusalem. The first half of AD69 saw the Romans reconquering most of Judaea; but during those same months, the heart of the Empire was shattered by a cruel civil war, which saw the ascendancy and fall of three monarchs, but this did not affect the military operations taking place in Judaea. At a certain point, however, Vespasian was hailed Emperor by the legions under his command: as a result, having decided to end the civil war by assuming the imperial title for himself, he went to Rome and left his son Titus to finish the war against the Jews.

A Roman optio with crested helmet.
(Legio XI C.P.F. Hispaniensis)

Titus advanced through the hill country surrounding Jerusalem with great determination, killing any individual who tried to stop his march; as a result, an immense wave of Jewish refugees went to the capital of Judaea in search of shelter. The Romans completed their surrounding of Jerusalem and initiated a long siege, which was particularly difficult. The legionaries were unable to breach the city's strong defences and were forced to dig a trench around the circumference of the enemy walls, as well as to build an external wall that was as high as Jerusalem's walls. Later, after having built massive ramparts, the legionaries were finally able to assault the enemy positions in a direct way. The resistance of the Jews was desperate, since almost all the civilians trapped in the city took up arms against the foreigners. In the summer of AD70, after seven

A Roman Optio drinking from his canteen, made of leather. (Legio XI C.P.F. Hispaniensis)

months of siege, Titus used the collapse of some sections of Jerusalem's walls to breach the city and to burn it to the ground. The Temple, one of the last bastions defended by the Zealots, was completely destroyed just as all the other buildings. Thousands of captured Jews became slaves and the Temple's treasures, including the Menorah and the Table of the Bread of God's Presence, were transported to Rome in order to be paraded during Titus's triumph. The fall of Jerusalem, however, did not mark the end of the Great Jewish Revolt. In fact, it continued until AD73 when all the remaining rebel strongholds were conquered by the Romans, including the strongest one of Masada, which fell only after an epic siege that ended with the suicide of all the remaining Jewish defenders. During the following decades, Judaea rose up in revolt on several occasions, most notably during AD115–117 and later in AD132–136 and the Romans were always able to crush these new national revolts. After AD136, thousands of Jews left Judaea for other areas of the Mediterranean, thus initiating the long and complex historical process known as Jewish Diaspora.

The Parthian Wars

The greatest enemy of the Romans in the Middle East was the Parthian Empire, a major military power of Asia that had combat capabilities comparable to those of the Roman Empire. The Parthians, originally a nomadic people living in northeastern Iran, had gradually created an empire in the Middle East under the guidance of their ambitious Arsacid Dynasty. By taking advantage of the Seleucid Empire's great internal and external difficulties, they had conquered the whole territory of modern day Iran and had later reached the borders of Syria (the heartland of the Seleucids) after having established themselves in a stable way in Iraq.

The Parthian military forces consisted entirely of cavalry: mounted archers armed with deadly composite bows and heavy cavalrymen clothed with armour. On several occasions, the Seleucids were defeated by this combination of light and heavy cavalry, which was extremely effective tactically. The Parthians knew how to fight for long periods in desert areas and had great logistical organisation. Their archers were superb skirmishers, who could fire their arrows with great accuracy including on horseback while fleeing from target (this being the famous 'Parthian shot'). The Parthian heavy cavalrymen or cataphracts, were equipped with long spears and rode armoured horses: when charging at full gallop, their power was impressive.

When the Seleucid Empire crumbled, the Romans occupied Syria and its other western regions, while the Parthians occupied Iraq and its other eastern regions. It soon became apparent, however, that the two emerging powers of the Middle East would soon fight each other for dominance of the whole region: the Romans wanted Iraq, the Parthians wanted Syria. The first major Roman–Parthian War took place in 53BC and ended with disastrous defeat for the Roman Republic, since a large Roman army commanded by Crassus was completely wiped out at the famous Battle of Carrhae. The Romans had underestimated the military capabilities of the Parthians and were soundly defeated.

During the Roman civil wars of the following years, the Parthians first sided with Pompey against Caesar and then with Brutus and Cassius (the murderers of Caesar) against Mark Anthony and Augustus. In 36BC, Mark Anthony led an inconclusive military campaign against the Parthians, which ended without achieving a positive result for Rome. After becoming princeps, Augustus tried to solve the Parthian problem with diplomacy since he had no intention of fighting a new war against the Arsacids. He obtained the restitution of several Roman standards that had been captured by the Parthians at Carrhae and transformed Armenia into a client kingdom of the Roman Empire. Because of its geographical position, in the

A Roman Optio with scale armour. (Legio XIIII Gemina)

middle between Parthian northwestern Iran and Roman Anatolia, Armenia soon became extremely important for Rome since it acted as a buffer zone between the two rival empires.

In AD58, following many years of increasing tension, a new conflict broke out between the Romans and the Parthians. The war began as a Roman military intervention into an ongoing Armenian civil conflict, which saw the clash between a weak pro-Roman faction and a strong pro-Parthian faction. The Roman military forces, including three legions (*III Gallica*, *IV Scythica* and *VI Ferrata*), initially obtained a series of easy successes and occupied the two capitals of Armenia: Artaxata and Tigranocerta.

In AD61, the Parthians mounted a massive counter-offensive with substantial military forces. They attempted to besiege Tigranocerta, but with no success, Later, however, they surrounded a large Roman military force during the Battle of Rhandeia and obliged it to surrender. In AD63, the hostilities came to an end with a compromise: the pretender supported by the Parthians obtained the Armenian throne, but he had to swear loyalty towards Rome as a vassal king of the Empire.

During the following decades, both the Romans and the Parthians continued to plot in Armenia in order to augment their political influence in that kingdom. In AD113, however, Emperor Trajan (AD53–117) – who had already conquered Dacia and who had demonstrated he was the greatest military leader in the history of the Empire – organised a large-scale invasion of the Parthian territories with the

A Roman optio with crested helmet. (David Burns, Legio XIIII Gemina)

objective of conquering them. The *casus belli* for the new conflict was the fact that the Parthians had installed a new monarch on the Armenian throne who was favourable to them. Very soon, however, it became clear that the new war would have been a real 'struggle for life' for both empires. Trajan planned the campaign well and assembled a large invasion force comprising an impressive ten legions. First, he invaded Armenia, deposing the pro-Parthian king and annexing the realm to the Roman Empire as a new province. He then marched down towards the Taurus Mountains to occupy the area of northern Mesopotamia (modern Iraq) located between the Tigris and Euphrates rivers. The Parthians were unable to stop the invaders and had to abandon some of their major Mesopotamian cities.

After spending the winter of AD115–116 in Antioch, Trajan resumed his general offensive and divided his forces into two columns: the first crossed the Tigris and then moved towards southern Mesopotamia, while the second captured the city of Babylon. Having come to the narrow strip of land located between the Euphrates and the Tigris, Trajan re-united his military forces and marched on the ancient Seleucid capital of Seleucia. Later, without fighting a single major pitched battle, he also entered the Parthian capital of Ctesiphon. The Parthians had been surprised by the strategic intelligence and logistical mastery of the Romans, who supplied their advancing troops during the whole campaign and transported a fleet of warships on the Tigris and on the Euphrates. Trajan reached the Persian Gulf and occupied the territory of modern Kuwait.

For the first time in history, the Romans reached the Indian Ocean and could organise Mesopotamia as a province: the Roman Empire had reached its maximum territorial extension and had defeated the Parthians in a decisive way. The Roman success, however, did not last for long since the newly conquered territories were too vast to be garrisoned in a Roman way, and since they were inhabited by peoples hostile to Rome. As soon as Trajan left the Persian Gulf for Babylon, the Parthians launched a counter-offensive against Mesopotamia and Armenia from Iran. The Romans responded by reinforcing their positions in northern Mesopotamia, but were forced to abandon southern Mesopotamia. During this final phase of the war, Trajan's health started to fail; the Emperor died during the siege of an enemy fortified city, at Hatra, probably after suffering heat stroke.

With Trajan's death in AD117, the war between the Romans and the Parthians soon came to an end. The new Emperor, Hadrian (AD76–138), promptly reversed his predecessor's policy for the Middle East. He decided that the Euphrates River would mark the border between Roman

A Roman vexillifer wearing lorica hamata.
(Legio XIII Gemina)

A Roman vexillifer wearing lorica segmentata. (Marc Seriol (@marcmarkhus_photo), Legio II Traiana Fortis – Cohors I Barcinonum, Barcino Oriens)

and Parthian territories; as a result, the status quo of the region was restored and Mesopotamia was returned to the Parthians. Armenia remained a vassal kingdom of the Empire, albeit enjoying a greater degree of autonomy than previously.

In AD161, war between the Roman Empire and the Parthian Empire, broke out again over Armenia when a Parthian army defeated the Roman military forces in Armenia and ravaged part of Roman Syria. The Romans responded by organising an effective counter-attack, which ended with the defeat of the Parthians in Armenia and with the installation of a new monarch on its throne, who was favourable to Rome.

After having taken the initiative, in AD164, the Romans launched a new invasion of Mesopotamia with the objective of re-conquering the territories that had once been occupied by Trajan. They won two battles against the Parthians (at Dura-Europos and at Seleucia) and were later able to sack the Parthian capital of Ctesiphon in AD165. The outbreak of an epidemic, possibly of smallpox, ended the war when it seemed that the Romans were on the verge of victory. The status quo of the Middle East was restored, but the legionaries who had participated in the conflict took the epidemic to the other territories of the Empire, initiating the Antonine Plague.

During the history of the Empire, the Romans were never able to eliminate the Parthian presence in the Middle East. In AD195–197, Emperor Septimius Severus (AD145–211) invaded Mesopotamia, and in AD216–217 Emperor Caracalla (AD188–217) also attacked the Parthians. These two later conflicts, however, were not decisive. The Parthian Empire was finally destroyed in AD226, when a new Persian dynasty – that of the Sassanids – emerged in Iran and conquered the city of Ctesiphon after a brief but bloody conflict. The Sassanids organised a centralised state that was much stronger than the semi-nomadic Parthian one and as a result, during the later centuries of the Empire, they invaded the Roman territories on several occasions.

The Year of the Four Emperors

The Roman state experienced a single major civil war that took place during the crucial year AD69. That year saw the end of the Julio-Claudian Dynasty created by Augustus and the ascendancy of the new Flavian Dynasty founded by Vespasian. Since the beginning of the principate, Roman emperors had experienced problems with their succession; the imperial monarchy of Rome, in fact, was not organised according to hereditary principles as it was in other states of the Mediterranean. Beginning with Augustus, the Emperors usually 'adopted' an heir who was part of their family but who was not their son. In most cases, in fact, Emperors did not have direct heirs. This 'adoption' principle worked well for some of the early Emperors, but soon showed its limits. Since no official dynasty existed, both the Senate and the Praetorian Guard could influence the election of the new monarchs by using their political or military influence.

Court intrigues taking place in Rome started to determine the leadership of the Empire, something that was not accepted by the Roman elites living in the provinces and by the superior military officers commanding the legions on the field. These institutions also wanted to play an important role in the political processes that led to the appointment of the new emperors. Over time, the will of the legionaries became the most important factor behind the election of new monarchs, since most of Rome's power came from the weapons of the legions.

Considering that the Empire was still in its expansionist phase of its history, it needed a strong military leadership to continue flourishing; the legions knew this and always supported potential emperors who had great military experience. Very often the will of the legionaries was in contrast to that of the elite Praetorians, who were highly corrupted and usually supported candidates who promised them large sums of money. The Senate, instead, was favourable (on most occasions) to potential emperors who came from the aristocratic families of Rome and who wanted to limit the powers of the emperor's role. The conflicting interests became apparent when the Julio-Claudian Dynasty came to an end with Nero's death.

Nero governed Rome well during the first years of his reign, but later made a series of terrible mistakes. He was unable to deal with the complex defence of the imperial lands, increased taxation above all limits, persecuted members of the Senate who tried to limit his personal powers, wasted enormous sums of money for futile reasons, failed to crush the Great Jewish Revolt and – probably – also caused the Great Fire of Rome (AD64). After having agreed that Nero would have soon destroyed the Empire with his personal ambitions and diminished thinking, the members of the Senate started to plot against him. A first conspiracy, known as the Pisonian Conspiracy, attempted to remove Nero and to restore the Republic in AD65 but was discovered by the Emperor (who ordered the execution of many important personalities, including Seneca).

In AD68, the governor of Gallia Lugdunensis, Gaius Julius Vindex (AD25–68), rebelled against Nero's tax policy and started planning to substitute the Emperor with Servius Sulpicius Galba, the experienced governor of Hispania Tarraconensis. Vindex's revolt, however, soon failed since the legions stationed on the Germanic border – the most powerful ones of the Roman Empire – remained loyal to Nero and marched against Gallia Lugdunensis.

In the early summer of AD68, the commander of the Praetorian Guard, Nymphidius Sabinus (AD35–68), incited his men to transfer their loyalty from Nero to Galba and thus started to plot against the ruling Emperor. The Senate tried the Emperor and condemned him to death as a public enemy. Nero, abandoned by all his remaining supporters, committed

A Roman signifer wearing scale armour. (Legio XI C.P.F. Hispaniensis)

suicide before he could be executed. With Nero's death, the Julio-Claudian Dynasty came to an end and a period of complete anarchy began for the Roman Empire. Galba was invited to come to Rome as new emperor at the head of a single legion (the *VII Gemina*) and was welcomed by the Senate. He was a moderate but capable administrator, who had served the Empire for most of his life. He respected the traditional prerogatives of the Senate and was strongly determined to counter the abuses and the corruption that had been favoured by Nero during the last part of his reign.

Dissent in the legions

With Nero's death, however, the legions serving on the Germanic limes lost the opportunity to receive a generous donation of money for having crushed the rebellion of Vindex in Gaul. As a result, they did not accept Galba's rise to power. To prevent the outbreak of a military revolt on the Germanic border, Galba appointed Aulus Vitellius (AD15–69) as new commander of the legions. Soon, however, it became clear that the latter was determined to use his new position only to increase his personal power and not to support the ruling Emperor. Once in Rome, Galba cancelled all the laws that had been promulgated by Nero and refused to pay the Praetorians with the reward that had been promised to them by their commander Nymphidius. With the beginning of the new year, AD69, the legions in Germania refused to swear allegiance and obedience to Galba. The soldiers, in fact, acclaimed their commander Vitellius as new Emperor. Upon hearing the news coming from the north, Galba decided to adopt a young senator, Lucius Calpurnius Piso Licinianus (AD38–69), as his successor in the hope of retaining the support of the Senate. His choice for the succession, however, offended the most influential and ambitious aristocrat of Rome who had hoped to be Galba's successor: Marcus Salvius Otho (AD32–69). The latter bribed the Praetorian Guard and organised a military coup. Both Galba and Lucius were killed in the streets of Rome, together with their few supporters.

At this point, there were two Emperors ruling at the same time: Otho who was supported by the Senate and by the Praetorian Guard in Rome; Vitellius who was supported by the legions in Germania. Very soon hostilities between the two monarchs became unavoidable, since neither intended to resign. Vitellius was backed by some of the finest legions of the Roman Army, which consisted of battle-hardened veterans of the Germanic Wars. These soon marched on Italy and, after obtaining minor victories, finally met with Otho's military forces at the First Battle of Bedriacum. The invading army of Vitellius included *Legio V Alaudae, Legio XXI Rapax* and several vexillationes from other legions in addition to a strong contingent of Batavian auxiliaries. Otho's military forces included *Legio I Adiutrix, Legio XIII Gemina*, the Praetorian Guard and a force of gladiators. The First Battle of Bedriacum was particularly violent and ended with the complete victory of Vitellius, whose veterans were soldiers of excellent quality. Following the clash, what remained of Otho's army surrendered and took the oath of allegiance to Vitellius; Otho decided to commit suicide rather than continue a hopeless resistance. He had been Emperor for fewer than three months and had been the second Roman monarch, after Nero, to commit suicide. After making a triumphal entry into Rome, Vitellius was recognised as the new legitimate Emperor by the Senate.

Vitellius spent his days in Rome organising banquets and triumphal parades, which had enormous costs for the imperial treasury that was already on the verge of bankruptcy. The Emperor ordered the torture and execution of all citizens suspected of being against him, and began to rule as a tyrant. While Vitellius consolidated his power in the capital of the Empire, the legions fighting in Judaea to crush the Great Jewish Revolt acclaimed their commander Vespasian as the new Emperor.

After having obtained the support of the governors of Egypt and Syria, Vespasian soon organised an expeditionary corps to invade Italy and remove Vitellius from the throne. Before his troops could reach the heart of the Empire, however, the legions stationed in the Danubian provinces of Raetia and Moesia rose up in revolt against Vitellius. They soon marched on Italy after acclaiming Vespasian as Emperor and fought against the supporters of Vitellius in what became known as the Second Battle of Bedriacum. The army sustaining Vespasian comprised *Legio III Gallica*, *Legio VII Claudia*, *Legio VIII Augusta*, *Legio VII Galbiana* and *Legio XIII Gemina*; the military forces supporting Vitellius comprised *Legio I Italica*, *Legio IV Macedonica*, *Legio V Alaudae*, *Legio XXI Rapax* and *Legio XXII Primigenia*. The Second Battle of

Above left: A Roman signifer wearing scale armour. (Legio VI Victrix Cohors II Cimbria)

Above right: A Roman aquilifer carrying the small round shield that was typical of standard-bearers. (Marc Seriol (@marcmarkhus_photo), Legio II Traiana Fortis – Cohors I Barcinonum, Barcino Oriens)

Bedriacum was a disaster for Vitellius, who was later captured in Rome when Vespasian's forces occupied the city (he was executed shortly afterwards). As a result of these events, the dramatic Year of the Four Emperors ended with the ascendancy to the throne of Vespasian and with the creation of the new Flavian Dynasty; which ruled Rome until AD96.

Dacia

Around AD70, the Dacian king Scorilo died and was succeeded by Duras, an ambitious warlord who was determined to continue the anti-Roman policy of his predecessor. He started to re-build the military power of the Dacians and did his best to unify the various tribes of his people. In AD85, Duras assembled a large army and invaded the Roman province of Moesia from the north: the attack soon had great success, since the Romans were unable to stop the Dacians. An entire legion was annihilated by the invaders and the same governor of Moesia, Oppius Sabinus, was decapitated during the major clash of this campaign.

At this point, it became clear in Rome that there was a strong enemy leader in that sector of the Danubian frontier, who had the needed military resources to occupy Moesia. Emperor Domitian (AD51–96), second son of Vespasian and last monarch of the Flavian Dynasty, decided to act very quickly. He went to Moesia at the head of a large military force and moved three legions to the territory of the menaced province. When the Roman troops arrived, however, the Dacians avoided a direct confrontation with the superior forces of the enemy. The war continued for several months, but without any major change in the strategic situation. On one occasion, the Romans were ambushed and defeated by the Dacians, but the Dacians suffered some defeats.

In AD86/87, Duras abdicated in favour of a younger Dacian war leader, named Decebalus (d.AD106). Decebalus had been the main military commander of the Dacians since the beginning of the conflict with Rome and had shown great personal capabilities on several occasions. The new king realised that there was no point continuing the war with Rome, at least for the moment, since the Roman garrison of Moesia was too strong to be defeated. Once Emperor Domitian had left the theatre of operations, a peace treaty was concluded between the Dacians and the Roman Empire. According to it, Decebalus returned the many Roman prisoners who had been captured by

Detail of the helmet of a Roman aquilifer, covered with the head of a wolf. (Marc Seriol (@marcmarkhus_photo), Legio II Traiana Fortis – Cohors I Barcinonum, Barcino Oriens)

the Dacians and promised not to attack Moesia in the future. In exchange, he obtained Roman military assistance to build new fortifications in his kingdom and an annual subsidy of eight million sesterces.

Once the new fortifications were built with the help of the Romans, Decebalus could complete the re-unification of Dacia initiated by his predecessor and transform it into a centralised state. With the large amounts of money sent every year by Rome, he could enlarge and re-equip his military forces, transforming them into a very efficient combat machine. Decebalus soon proved to be an excellent king, both from a military point of view but also from an economical and administrative one. He started to exploit, in a methodical way, all the great minerary resources of his kingdom in order to produce larger amounts of precious metals. He reorganised the structures of his state to exert a better control over the peripheral areas of Dacia (where the local tribes were all brought under his influence). In addition, Decebalus also ordered the creation of a fortified capital for his realm, known as Sarmizegetusa. It soon became the most important political and religious centre of Dacia, sited on top of a mountain, at an altitude of 1,200m, and in the centre of the Dacian Kingdom.

Sarmizegetusa was extremely easy to defend. It was made up of six connected citadels and was part of a larger system of fortifications. Sarmizegetusa had already functioned as the capital of the Dacians during the reign of Burebista, but after his death, it had lost its special status. In addition, before the rise of Decebalus, it had never been fortified or enlarged.

During the decade that followed the end of the hostilities between Dacia and Rome, Decebalus transformed his kingdom into one of the most flourishing states of Antiquity. He controlled a large portion of European territory, corresponding to modern Romania, and could field an impressive army of more than 200,000 warriors. Before long, the Romans understood that signing a peace treaty with the Dacians had been a mistake, especially because the conditions offered to Decebalus had been very generous. Sooner or later, the Dacians would attack Moesia and the Empire had to be ready to react.

The Romans did not have good memories of their previous war fought against the Dacians: the defeat of Oppius Sabinus and the humiliation suffered by the legions, in fact, had not been forgotten. Decebalus had been responsible for that Roman failure, when he was still a military commander. At that time, his name was Diurpaneus, since he received the new name 'Decebalus' (meaning 'the Brave') after defeating and killing Oppius Sabinus.

In AD97, Trajan became Emperor and started to deal with the most important military issues of his state. He considered Dacia as a substantial threat for the stability of the Empire, since it was too large to be considered as a standard client state of Rome. In AD101, after obtaining the Senate's official blessing, Trajan started military preparations for the invasion of Dacia. By conquering that country, the Romans would

A Roman cornicen with lorica hamata.
(Legio VI Victrix Cohors II Cimbria)

Detail of the parade mask worn by a cornicen. (Legio VI Victrix Cohors II Cimbria)

have stabilised their northern borders of Empire on the Danube. In addition, they would have obtained access to the vast natural resources of a region that had never been fully explored. Culturally, Dacia was a kingdom located on the edges of the known world: the Romans did not have a proper knowledge of it and considered the Dacians as one of the many barbarian peoples living on their frontiers.

In any case, Trajan prepared his invasion with great attention and assembled large military forces to conduct it. The Dacians also prepared, under the intelligent guidance of Decebalus. The king made great efforts to conclude an important military alliance with the Sarmatians. These were a nomadic people from the steppes of Central Asia, who had gradually driven out the Scythians from the plains of modern day southern Russia and Ukraine during the third century BC. Like their Scythian enemies, the Sarmatians spent most of their lives on horseback. Their armies were made up of mounted archers equipped with composite bows, and of cataphracts (cavalrymen with full armour) armed with heavy spears. Over time, the Sarmatians started to move south and practically eliminated the Scythians; occupying all the territories and also establishing themselves north of Dacia. Decebalus was able to conclude a solid alliance with these steppe warriors and also the powerful tribe of the Roxolani, who provided him with large contingents of top quality cavalry.

Trajan reached the Roman province of Moesia in the spring of AD101. He was at the head of one of the largest military forces ever assembled in the history of Rome. It comprised the following units: 15 legions, 10 vexillationes (detachments) of legions and 89 corps of auxiliaries. The legions were the *I Adiutrix, I Italica, I Minervia, II Adiutrix, II Traiana Fortis, III Flavia, V Macedonica, VII Claudia, X Gemina, XI Claudia Pia Fidelis, XIII Gemina, XIV Gemina Martia Victrix, XV Apollinaris, XXI Rapax* and *XXX Ulpia Victrix*. The vexillationes came from the following legions: *II Augusta, III Augusta, III Gallica, IV Scythica, VI Ferrata, VII Gemina, IX Hispana, XII Fulminata, XX Valeria Victrix* and *XXII Primigenia*. The corps of auxilia comprised 21 alae of cavalry, 33 *cohortes equitatae* (mixed infantry and cavalry), 25 *cohortes peditatae* (infantry) and 10 *cohortes sagittariae* (archers). In total, Trajan could count on 75,000 legionaries and 55,000 auxiliaries. These were supplemented by another 20,000 auxilia soldiers who were transferred to Moesia specifically for this campaign, (the other 55,000 were already stationed along the Danube before the outbreak of the hostilities). The Emperor also brought the famous Praetorian Guard with him.

The Roman invasion force was divided into two large columns, which crossed the Danube on two pontoon bridges that were built with the ships of the Roman river fleet. Trajan's war plan was simple: he wanted to cross southern Dacia as soon as possible, ravaging all the enemy settlements encountered along the way, in order to rapidly reach the 'Iron Gates'. These were a narrow mountain pass located west of the Dacian capital. Capturing them was the only way to approach Sarmizegetusa and to enter the main fortified system of the Dacians.

During the first phase of the campaign, Decebalus avoided a direct confrontation with the Romans and employed 'scorched earth' tactics. All food reserves of southern Dacia were moved north or destroyed before the Romans could capture them. The Dacians retreated towards the heart of their territories and obliged the Roman Army to move across densely forested areas.

The Romans knew very little of the interior regions of Dacia, and over time their supply lines became increasingly stretched to the point of being dangerously exposed to Dacian incursions. Trajan continued his advance very slowly, to avoid possible ambushes and to consolidate his presence on the enemy territory. While moving north, the Romans built camps, roads and bridges: so that they would be able to put up strong resistance in southern Dacia in case of defeat. When the Romans reached the Iron Gates, however, Decebalus decided that the time had come to fight a large pitched battle. In what became known as the Second Battle of Tapae, which was extremely hard for both sides, the Romans were able to repulse the Dacian assaults, but only after suffering severe losses. Decebalus also lost many warriors, but he had time and resources to replace them.

At this point of the war, the Dacians retreated behind the Iron Gates and inside their main fortified system. The Romans built a massive winter camp not far from the entrance to the mountain pass. During the first months of AD102, with the Roman Army blocked outside the Iron Gates, Decebalus decided to act on another front with the objective of diverting Trajan's attention.

Together with his Roxolani allies, Decebalus attacked Moesia from the north at the head of a massive military force. The Roman garrison of Moesia had great difficulties in containing Decebalus's invasion, but with the arrival of substantial reinforcements sent by Trajan, the general situation improved. The Dacians and Roxolani made the fatal mistake of separating their forces, and this enabled the Romans to react in a more effective way. Both the

A Roman cornicen with lorica segmentata. (Marc Seriol (@marcmarkhus_photo), Legio II Traiana Fortis – Cohors I Barcinonum, Barcino Oriens)

Dacians and the Roxolani were defeated, thus obliging Decebalus to abandon his plans for the opening of a second front. After these events, Trajan resumed his offensive in Dacia, but only after having reorganised his army into three separate columns. These would have attacked the Dacian fortifications from three different directions, having as their final objective the conquest of Decebalus's capital.

With great difficulty, the Romans were able to conquer several Dacian fortifications and started to encircle Sarmizegetusa. At this point, in order to buy time and re-organise his defences, Decebalus sent emissaries to Trajan with offers of peace. The Romans responded by proposing very harsh conditions, which were not acceptable by Decebalus. Military operations resumed very soon. During the following weeks, the Romans besieged and conquered all the Dacian fortifications located around Sarmizegetusa.

The Dacian Army attacked the Roman troops while they were besieging the last stronghold, but this offensive was repulsed with heavy losses for the attackers. After these events, the way to the Dacian capital became open for the Romans. Decebalus's army had been defeated on the open field and there were no fortifications left in Dacian hands. The Romans, however, were extremely tired: they had suffered significant losses since the beginning of the war and were far from their provinces.

The siege of Sarmizegetusa would have lasted for months, since the city was heavily fortified and built on top of a mountain; Trajan could not sustain a new campaign in a hostile land, especially without proper supplies and fresh reinforcements. As a result, he decided to terminate hostilities and to conclude a peace treaty with Decebalus. The conditions imposed by the Romans on the Dacians were very harsh: Decebalus had to accept the presence of some Roman garrisons on his territory and was forced to give up all his weapons. In addition, he was asked to destroy all the fortifications of his realm and to cede part of

Above left: A Roman cornicen. (Marc Seriol (@marcmarkhus_photo), Legio II Traiana Fortis – Cohors I Barcinonum, Barcino Oriens)

Above right: A Roman tesserarius, with the small block of wood on which watchwords (secret codes known among the legion) were written. (Marc Seriol (@marcmarkhus_photo), Legio II Traiana Fortis – Cohors I Barcinonum, Barcino Oriens)

his southern territories to the Roman Empire. Finally, the Dacians were also required to accept Rome's protection and thus transform their kingdom into a client state of the Empire. Decebalus had no choice but to accept these humiliating conditions, in order to gain some time and prepare for a new war against Rome.

In AD102, the military operations thus came to an end, at least for the moment. Soon after Trajan abandoned the theatre of operations with most of his troops, Decebalus started organising a new war. He re-equipped his whole army with new weapons and re-built the fortifications located around his capital that had been destroyed by the Romans. He attacked the Iazyges, a Sarmatian tribe that was allied with Rome, and ordered the execution of all those Dacian nobles who were in favour of respecting the peace treaty signed with the Romans.

During AD105, following these events, Trajan decided to respond and moved again at the head of his forces towards Moesia. When the Roman Army arrived on the Danube, Trajan learned that his garrisons in Dacia had been massacred and that Decebalus was already waiting for him north of the river. Many Roman fortresses on the Danubian limes had been attacked and occupied by the Dacians; as a result, Trajan had to spend the entire summer of AD105 reconquering the fortified positions that had recently been lost. During this early phase of the new war, the Romans realised it had been a mistake not to besiege Sarmizegetusa during the previous conflict. It was clear by now, that the only way to secure the Danubian frontier was to defeat Dacia and annexe it to the Empire.

Decebalus was too intelligent and too ambitious to be a vassal king, so it was essential to conclude the new war with his capture or with his execution. Knowing that conquering the whole territory of the Dacian Kingdom would have been very difficult, Trajan remained south of the Danube until AD106. During this period, he reorganised the defences of Moesia and assembled more military forces for the upcoming invasion. During these crucial months of preparation, the Romans constructed a permanent bridge across the Danube. Because of this, the Roman Army would have been able to move north at a moment's notice and would have received supplies and reinforcements much more easily than in the previous war.

The Dacians were particularly impressed by the building of this bridge, which was absolutely incredible for the standards of the time. However, they could do very little to destroy it since it was defended by a substantial number of Roman troops. When it became clear that Trajan would have crossed the Danube at the head of a massive invasion force, the main allies of Decebalus abandoned the Dacians to their destiny. The Roxolani and the Bastarnae, who had made up a consistent portion of Decebalus's military forces during the previous conflict, decided to change their attitude and proclaimed their neutrality. This way, they tried to avoid the bloody vengeance of the Romans.

While these events took place north of the Danube, Trajan did his best to kindle military support from the Iazyges and from the Germanic tribes that were settled on Dacia's northwestern border. He wanted to be sure that these peoples would not collaborate with Decebalus and he hoped that they could attack Dacia from other directions in order to open other fronts on the Dacians' frontiers. After realising that Decebalus had been abandoned by his allies, the Romans initiated a new invasion of Dacia. The Roman Army was divided into two large columns: one to attack the fortifications of Sarmizegetusa from the west (forcing the Iron Gates), and the other to attack from the east. The Roman advance was very difficult and slow because the Dacian warriors put up strong resistance and defended their homeland to the last man. The fortifications re-built by Decebalus after AD102, however, proved to be of inferior quality compared to those that had been constructed several years before with the assistance of Roman military engineers. As a result, they were all conquered by the advancing Romans and their garrisons were either destroyed or captured.

By the end of summer of AD106, all the Dacian fortified positions located around Sarmizegetusa had been conquered and the two Roman invading columns could join their forces. Trajan was now ready to besiege the Dacian capital with his superior forces, but he had to act very rapidly. Winter was coming and it would have been impossible for his soldiers to conduct a long siege on high ground with cold temperatures. The

A Roman legionary light infantryman armed with javelins. (Legio VI Victrix Cohors II Cimbria)

Emperor, however, was determined not to make the same mistake for a second time: Sarmizegetusa had to be taken and Decebalus had to be neutralised, at all costs. The siege of the Dacian capital did not last for long: the fighting between the defenders and the attackers was brutal, but the Dacians had lost most of their best warriors during the previous combat operations and its population was exhausted. When the Roman soldiers entered the city, understanding that all was lost, many Dacian leaders committed suicide in order to avoid capture. Decebalus, instead, did not consider the fall of his capital as a mortal blow: he abandoned Sarmizegetusa before the end of the siege and moved north.

The Dacian king knew the territory of his realm well and had a clear idea how to continue resistance against the invaders. He moved to the densely forested areas of the Carpathians, where he raised new military forces and started planning a new campaign: this was to be conducted with guerrilla methods, since the Romans were by now too numerous to be faced in a pitched battle. The northern portion of Dacia was the wildest of Decebalus's kingdom: it was located quite far from the Danube and did not have proper roads. Potentially, the Dacians could resist for years in the Carpathians; Decebalus, in fact, hoped that a long resistance would have given him enough time to form new alliances with the Sarmatians and to raise more warriors.

After conquering Sarmizegetusa, Trajan formed a special column of elite Roman troops and sent it to the north with orders to capture or kill Decebalus. In that moment, the Dacian king was no longer a menace to the stability of the Empire, since he commanded just a few retainers. If he was not neutralised, however, he could have reassumed control over his kingdom during the future by organising a revolt of the Dacians against the Romans. The troops sent to northern Dacia had great difficulties in finding the enemy and had to advance very slowly: the local terrain was perfect for ambushes and they had no idea of its extent. In this phase of the war, the Dacians obtained minor successes, usually by attacking small Roman parties with hit-and-run tactics. Time was running out, however, and Decebalus had not yet been able to conclude a new alliance with the Sarmatians. His military resources remained limited and the Romans were gradually conquering northern Dacia. In the end, during one of the many skirmishes that took place in this final phase of the conflict, Decebalus and his retainers were reached by a Roman cavalry unit made up of auxiliaries. Surrounded by enemy soldiers, the great Dacian king decided to commit suicide to avoid capture. All the other leaders who were with him did the same and so the Romans could capture only a few Dacians. The head of Decebalus was taken to Trajan a few days later.

With the death of the Dacian king, the war came to an end: the Romans continued to fight for some months in Dacia, but only to put down minor revolts erupting on a local basis. By the end of AD106, the

Above left: A Roman legionary light infantryman, with sun hat and cingulum militia, the traditional Roman military belt. (Marc Seriol (@marcmarkhus_photo), Legio II Traiana Fortis – Cohors I Barcinonum, Barcino Oriens)

Above right: A Roman auxiliary infantryman with lorica hamata and oval shield. (Marc Seriol (@marcmarkhus_photo), Legio II Traiana Fortis – Cohors I Barcinonum, Barcino Oriens) former Dacian kingdom had been transformed into a Roman province. The Empire now comprised a large portion of territory located north of the Danube. The fall of the Dacian kingdom, however, did not mark the end of the Dacians as an independent people. Many of them, abandoned their home territories after the Roman conquest and moved north in search of new lands to live according to their own traditions. These people soon became known as 'Free Dacians', to distinguish them from the majority of the Dacians who had remained in their homeland under Roman rule.

The Free Dacians settled on territories already inhabited by peoples of Dacian stock and where the presence of the Sarmatians was quite significant. These areas today form part of southwestern Ukraine, Moldavia and Bessarabia (a region of Romania). The two main communities of Dacia already inhabiting these locations were the Costoboci and the Carpi. The former lived between the Carpathians and the Dniestr River; the latter between the Siret River and the Prut River. Ethnically they were Dacians, but they had never been included in the Dacian kingdom. Down the decades, they had been strongly influenced by the material culture of the Sarmatians and so were quite different from the Dacians.

Soon, the Costoboci and the Carpi absorbed the refugees coming from the Dacian kingdom into their own communities and became part of the Free Dacians. Once absorbed, the Free Dacians started to launch frequent raids against Roman Dacia: they never accepted the loss of their homeland and always hoped that the Romans would abandon it. The northern borders of Dacia were particularly difficult to defend for the Romans: there was no great river to use as a natural barrier and the frontier was too long to be defended with the construction of a wall.

The Batavian Revolt

Up to the outbreak of the Marcomannic Wars in ad166, the military situation on the frontier between the Empire and the Germani remained relatively quiet. Some Germanic tribes occasionally launched incursions across the Rhine, to which the Romans responded with punitive raids or brief campaigns. In general, however, very little changed politically and so the Romans could consolidate their frontiers on the Rhine. Only one event, taking place during ad69–70, saw the Romans fighting on a large scale against the Germani: the so-called Revolt of the Batavi.

As we have seen, the Batavi had been conquered by the Romans several decades before and had since been Rome's most loyal allies in central Europe. They had participated in the campaign of Germanicus against Arminius, to the invasion of Britain and had even provided a chosen bodyguard to the Emperors in Rome. In ad69, Nero, who was particularly loved by the Batavi because he greatly appreciated their loyalty, committed suicide; following this event, the Roman Empire entered into a brief but bloody new historical phase that was characterised by the outbreak of a civil war.

As previously discussed, ad69 is known as the Year of the Four Emperors, since it saw the rapid rise and fall of three new rulers who tried to take the place of Nero. Ultimately, this internal struggle ended with the ascendancy of Vespasian as Emperor and the installation of a new dynasty, the Flavian one, which replaced the original Imperial family created by Augustus. The Year of the Four Emperors saw the disbanding of the *Germani corporis custodes* in Rome and a rapid change in the relationship that existed between the Romans and the Batavi. When Nero died, however, malcontent started to spread among the Batavi who considered the disbanding of the Germani corporis custodes as an insult to their national pride. At that time, the Batavi provided eight *cohortes* of infantry and one *ala* of cavalry to the Roman Army, which were attached to the XIV Legion. When Vitellius (one of the pretenders to Nero's throne) tried to raise more troops from the Batavi for his military campaigns, the Germani rose up in open revolt and attacked the legion to which their units were attached.

The Batavian Revolt that took place in the modern Netherlands was guided by a hereditary prince who had served for a long time and with distinction in the Roman Army. This was Gaius Julius Civilis, who soon gained the upper hand in the campaign since most of the best Roman troops were fighting in the ongoing civil war. The Batavian leader was a man of experience, since he had served as overall commander of the Batavian auxiliaries during the invasion of Britain. In September AD69, the Batavi attacked the fortified camps of the two Roman legions that had been sent against them and later obtained a brilliant victory: the two legions, in fact, were forced to surrender after a long siege. The Romans were permitted to leave their camp according to the terms of a truce that was signed with Gaius Julius Civilis, but soon after having abandoned their positions they were ambushed by the Batavi who were waiting for them. Both legions were destroyed and most of their commanders were enslaved. After this incredible success, the Batavi were joined by several Gallic and Germanic tribes: the Batavian Revolt was expanding, but by now the Year of the Four Emperors was over and Vespasian had obtained a stable control over the Empire. The new Emperor assembled a very large army to crush the rebellion, which comprised a total of eight legions. The homeland of the Batavi was invaded by the Romans, who had to fight against the Germani also at sea since they had a strong fleet operating in the English Channel. In the end, the Batavi were crushed and had to accept peace on humiliating terms: they lost any form of independence and their home territory started to be garrisoned on a permanent basis by the Roman Army. The events of the Batavian Revolt, albeit the result of an ongoing Roman civil war, had shown once again the great military potential of the Germani. A single Germanic tribe had been able to defeat two legions and to regain, albeit temporarily, its former autonomy.

Above left: A Roman auxiliary infantryman. Note the protective leather cover of the oval shield. (David Burns, Legio XIIII Gemina)

Above right: A Roman auxiliary infantryman. (Legio VI Victrix Cohors II Cimbria)

Chapter 5

From the Marcomannic Wars to the Crisis of the Third Century, AD166–235

The Marcomannic Wars, also known as the *Bellum Germanicum et Sarmaticum*, were a series of conflicts that began in AD166 and lasted for 14 years. They included the clash between the Roman Army of Marcus Aurelius and the Germanic tribes of the Marcomanni and Quadi (supported by the Sarmatians, a nomadic people coming from the steppes of central Asia). The struggle took place on the Danubian frontier of the Roman Empire, along which the Marcomanni had lived since the days of Maroboduus. Around AD160, in the vast plains of modern day Eastern Europe, an important historical process known as Great Migrations began. It happened, at least initially, far from the eyes of Rome and could ultimately have had terrible consequences for the destiny of the Empire. Coming under pressure from the war-like 'steppe peoples' that were migrating across central Asia, the Germanic tribes living in eastern Europe, such as the Goths, started to move west in search of new lands on which to live.

The steppe peoples were nomadic communities with strong military traditions: they had superior cavalry forces and thanks to the use of a deadly weapon such as the composite bow, they were able to easily defeat the eastern Germani and obliged the latter to abandon their home territories. To avoid destruction and retain their beloved freedom, the eastern Germani invaded the lands of the western Germani; as a result, the latter were forced to cross the borders of the Roman Empire in the hope of finding new lands on which to settle. Within a few years, the Romans started to feel increasing pressure on their frontiers, and that was just the beginning of a much larger historical process that would ultimately cause the fall of the Empire.

A Roman auxiliary infantryman wearing Celtic winter clothing. (Legio VI Victrix Cohors II Cimbria)

Unrest in the Empire

Until AD160, relations between the Marcomanni and the Romans had been quite positive. No significant conflicts had ever broken out along the western portion of the Danubian limes. The pressure of the Goths and of other eastern tribes, however, changed this situation. The Marcomanni tried to cross the borders of the Empire with the objective of establishing themselves in Pannonia (present-day Hungary) and in the other Roman provinces that were located south of the Danube. In AD162, the Germanic tribes of the Chatti and of the Chauci moved against the western sector of the Danubian limes, and were repulsed only after three years of incursions. In AD166, a group of Langobards invaded Pannonia for a short period before being crushed. These early incursions were the first beginnings of a new historical process, which the Romans had not yet understood.

The Marcomanni, with their leader Ballomar (AD140–170), initially tried to mediate between the Romans and the other Germanic tribes. Quite soon, however, the eastern part of the Danubian border came under attack. Dacia, one of the largest and most recently conquered of the Roman provinces, was invaded by the Vandals and by the Sarmatians who were able to obtain a clear victory over the local military garrisons. All these events took place in a crucial year, AD166, which saw the emergence of another great problem for the Roman Empire. In that year, the legionaries returning from a campaign that had been fought against the Parthians in Mesopotamia brought with them, into the borders of the Empire, a new disease that later became known as Antonine Plague (for the name of the dynasty that was ruling Rome at that time). In a few months, the plague killed eight million Roman citizens/subjects and greatly weakened the Empire. It happened at a time that was already very complex for the stability of the Roman state.

Marcus Aurelius reacted against the incursions of the Germanic tribes only in AD168, when the plague ended. The economy of the Empire was in a critical situation and the Roman Army was short of men. The Emperor established his headquarters at Aquileia, in northeastern Italy, and moved against the invaders of Pannonia. Those who had initially been repulsed, had later been joined by the Marcomanni and had returned to the Roman province during the difficult months of the Antonine Plague. When Marcus Aurelius reached Pannonia with his military forces, the Marcomanni preferred to cross the Danube out of the Roman Empire and thus avoided a direct confrontation with the Romans. In AD169, the Emperor turned his attention to Dacia and fought against the Sarmatians, obtaining little result. At the same time, the Costoboci entered the Balkans from the north and raided them as far as Greece. The Marcomanni, seeing an opportunity, also crossed the Danube and invaded Pannonia with a large number of warriors. They had just formed a strong military alliance with the Quadi, another important Germanic tribe.

A Roman auxiliary infantryman wearing Celtic winter clothing. (Legio VI Victrix Cohors II Cimbria)

In AD170, at the Battle of Carnutum, the Roman military forces in Pannonia were utterly defeated by the Germani and suffered more than 20,000 casualties. There are very few details about this clash, but what we know for sure is that it was a major setback for Marcus Aurelius. After obtaining such a great and unexpected victory, the Germani ravaged Noricum (modern Austria) and moved south towards Italy. The same city of Aquileia was besieged, and for the first time after centuries the territory of the Italian peninsula was menaced by a foreign invader. By the end of AD171, however, Marcus Aurelius had been able to mount an effective counter-offensive against the Germani. He assembled all the military resources that were still available to him and relieved the besieged city of Aquileia. After a few months, the invaders were repulsed to the northern bank of the Danube. At this point, the Quadi signed a temporary peace treaty with Rome, while the Marcomanni continued to fight.

In AD172, Marcus Aurelius crossed the great river and attacked Ballomar's homeland. The Roman offensive had success and led to the defeat of the Marcomanni; the latter, however, still had significant military resources to deploy. In AD173, the Quadi attacked the Romans again, after breaking the peace treaty that they had signed with the Empire. During that same year, the Germanic tribes of the Chatti and of the Chauci launched some heavy raids against the Roman limes on the Rhine. Marcus Aurelius concentrated all his efforts against the Quadi since he considered them, correctly, as the most important menace to the stability of the Empire. In AD174, the Germanic tribe was defeated in a decisive way. The Quadi were forced to accept the presence of Roman military garrisons on their territory, had to surrender all the prisoners/hostages who were in their hands, and were obliged to provide some auxiliary contingents to the Roman Army.

A Germanic auxiliary in Roman service. (Legio VI Victrix Cohors II Cimbria)

At this point, the Romans turned against the Sarmatians and obtained some important successes over them on the plains of Pannonia. In AD175, the Sarmatians agreed to sign a peace treaty with Rome, according to which they were to contribute a total of 8,000 soldiers to the military forces of the Empire. Some 5,500 of these were sent to Britain, where they garrisoned the northern border and protected it from the incursions of the Celtic raiders coming from Scotland. Before Marcus Aurelius could consolidate the Roman military presence north of the Danube, an internal revolt (headed by the usurper Gaius Avidius Cassius [AD130–175]) broke out in the eastern provinces and prevented the Emperor from expanding the Roman territories along the northern frontiers.

In AD177, the Marcomanni and the Quadi revolted against the Romans and attacked the Imperial garrisons that were stationed on their territories. The Marcomanni were easily crushed by Marcus Aurelius, but the military operations against the Quadi took more time. Marcus Aurelius died in AD180 while the campaign against the Germani was still going on. The Romans, however, had already prevailed in two pitched battles. Marcus Aurelius's son and successor, Commodus, had little interest in continuing the campaigns against the Germani or in stabilising the northern frontiers of the Empire. As a result, he concluded a superficial peace treaty with the Marcomanni and the Quadi without solving the problems of the frontiers. Since the end of the bloody Marcomannic Wars, which had tested the military resources of Rome to their limit, a total of 16 legions were to be stationed on the territories that bordered the Germanic lands. It is interesting to note that during the Marcomannic Wars some first communities of Germani were permitted to settle inside the borders of the Roman Empire. Their members were known as *laeti* and consisted of defeated barbarians who had received some lands in exchange for military service in the Roman Army. The Antonine Plague had depopulated large areas of the Empire, reducing a lot the recruiting potential of Rome. As a result, Marcus Aurelius made this early 'experiment' that was not a great success. The defeated Germanic warriors, in fact, soon rose up in revolt and even seized for a brief period the important city of Ravenna (one of the Roman Navy's main bases). The agreement granting land to a community of laeti might specify a once-and-for-all contribution of recruits or a fixed number of recruits required each year. The lands given to Germanic settlers were known as *terrae laeticae* and were not part of the standard provincial administration, since they were under direct control of a specific administrative officer known as *Praefectus Laetorum*. In addition to the Germani, a good number of Sarmatians were settled in Italy and Gaul as 'military colonists'.

The year of five emperors

The death of Marcus Aurelius in AD180 opened a phase of political instability for the Roman Empire since his son and successor, Commodus, soon proved to be unstable and unfit to rule the Roman state. The new monarch ordered several executions of aristocrats who had been loyal to his predecessor, and greatly reduced the influence of the Senate. He also spent large sums of money organising magnificent gladiatorial games in the Colosseum and did nothing to stabilise the military situation on the Germanic frontier. In AD193, three Roman nobles, fearing that they could be the next targets of Commodus's paranoia, organised the assassination of the Emperor who was strangled in the arena while fighting as a gladiator.

Pertinax (AD126–193), the experienced Proconsul of Africa, was chosen as the new monarch because of the positive relations he had with many members of the Senate. The Emperor, however, soon had to face a series of problems. He was accused of having organised Commodus's assassination, and also had to resolve the dramatic financial crisis that was affecting the Empire. When Pertinax refused to pay a large *donativum*

A Roman auxiliary archer from the eastern provinces of the Empire. (Legio XIII Gemina)

(gift) to the Praetorians, they rebelled against him since they wanted to keep the privileges that had been assigned to them by Commodus. After just three months of reign, Pertinax was assassinated by the Praetorians, who then auctioned the title of Emperor to the highest bidder.

Didius Julianus, who had already succeeded Pertinax as Proconsul of Africa, won the favour of the Praetorian Guard and became Emperor. Soon after his ascendancy, however, two important military leaders rose up in revolt against him: the Governor of Syria, Pescennius Niger (AD135–194), and the Governor of Pannonia, Septimius Severus. The latter marched with his legions on Rome and ordered the execution of Didius Julianus, who had acted as Emperor for less than two months. After these events, both Septimius Severus and Pescennius Niger were proclaimed emperors by their troops. AD193 later became known as the Year of the Five Emperors since it saw five different monarchs on the imperial throne: Commodus, Pertinax, Didius Julianus, Pescennius Niger and Septimius Severus. The last two fought each other in a bloody civil war that lasted until 194 and ended with the victory of Septimius Severus. However, Severus had to continue campaigning until 197 in order to pacify the Empire since one of his strongest political allies – Clodius Albinus – revolted against him. In AD197, Septimius Severus defeated, in a decisive way, Clodius Albinus at the Battle of Lugdunum and so remained as the sole monarch of Rome. He was a very capable and ambitious military leader, who wanted to expand the Roman territories both in Britain and in the Middle East.

Septimius Severus

In AD197, Septimius Severus went to the eastern provinces of his Empire at the head of a large army, with the intention of invading the Parthian lands. During the military preparations for the new campaign, he raised three new legions – I Parthica, II Parthica and III Parthica – that had an innovative character since they comprised two centuries of light infantry in each of their cohorts. The Emperor knew that his army needed some elite light infantry to face the Parthians on equal terms and so equipped some of his newly raised legionaries as javelinmen. After the end of the Parthian campaign, the II Parthica was transferred to Italy. No other Roman legion had even been garrisoned in the Italian Peninsula since the reign of Augustus. Septimius Severus disliked the Praetorians and wanted to replace them with his own loyal veterans of the II Parthica; the latter was specifically tasked with preventing internal rebellions as well as with acting as a 'strategic reserve' that could be sent to every corner of the Empire in case of need. In practice, it was the first form of comitatus praesentalis (centralised army).

Septimius Severus's Parthian campaign was quite successful: his troops raided the Parthian imperial city of Ctesiphon and annexed a portion of Mesopotamia to the Empire. Despite besieging it twice, however, the Emperor was unable to take the important Parthian stronghold of Hatra. Septimius Severus was born in Africa and so always paid special attention to the Roman territories located along the southern coastline of the Mediterranean. During AD202, for example, he significantly expanded the Roman presence in Africa at the expense of the war-like Garamantes and completely re-fortified the limes that separated the Roman provinces from the nomadic tribes living in the Sahara.

In AD208, Septimius Severus decided to invade Caledonia (present day Scotland) to secure the Roman presence in Britannia. For more than a century the Celtic tribes, later became known as Picts, living in Caledonia, had conducted raids and incursions across the Roman territories in northern Britannia. Even the construction of the famous Hadrian's Wall and of the lesser known Antonine Wall had not been enough to stop their military activities. The portion of Caledonia comprised between Hadrian's Wall in the south and the Antonine Wall in the north, which corresponded more or less to the Lowlands of Scotland, had been temporarily controlled by the Romans though control was weak. By AD208, however, the Antonine Wall needed urgent repairs and the Roman Army exerted its direct control only south of Hadrian's Wall. In that year, Septimius Severus arrived in Britannia at the head of 40,000 soldiers, with whom he rapidly marched north. He initiated a massive rebuilding project of Hadrian's Wall, which finally made the whole wall stone. Until that time, the western portion of the fortification had been made of turf and timber. The Romans re-occupied the whole area south of the Antonine Wall (which had been lost during the previous decades) and repaired it.

In AD209, Septimius Severus invaded the heartland of the Caledonians – the wild Highlands. Only the great general Agricola, one century before him, had attempted something similar. The Caledonians responded to the Roman invasion with guerrilla tactics, which caused severe human losses to their enemies. Septimius Severus employed very harsh methods against the Caledonians, devastating the lands of all the tribes that did not submit and killing thousands of civilians. When it seemed that the whole of Caledonia could be conquered by the Romans, during AD210, Septimius Severus became ill and retreated to York to recover. Here, in February AD211, he died. The Emperor was succeeded by his son Caracalla, who had participated with distinction in the war against the Caledonians. The new monarch, being obliged to consolidate his

Above left: **A Roman auxiliary archer from the eastern provinces of the Empire. (Marc Seriol (@marcmarkhus_photo), Legio II Traiana Fortis – Cohors I Barcinonum, Barcino Oriens)**

Above right: **A Roman auxiliary archer with bronze helmet. (Legio XI C.P.F. Hispaniensis)**

personal power in Rome, had no choice but to suspend the invasion and leave Britannia with most of his troops. The military events of AD208–210 had been the last Roman attempt to invade Caledonia: they greatly weakened the Celtic tribes of Scotland and secured the survival of Britannia, but soon after Caracalla's departure the Antonine Wall was abandoned again and the Lowlands were retaken by the Caledonians.

From Caracalla to Severus Alexander

Caracalla, like his father, spent large sums of money improving the combat capabilities of the Roman Army and continued to nurture the ambition of conquering the Parthian Empire. Unlike Septimius Severus, who believed that the Parthian mounted archers could be defeated by well-trained light infantry, Caracalla was convinced that the only way to limit the combat capabilities of the excellent Parthian cavalry was to re-equip the Roman legionaries as *phalangites* in imitation of Alexander the Great's Macedonian Army. For this purpose, he levied a new military unit from the Roman province of Macedonia – consisting of 16,000 men – whose members were trained and equipped as phalangites. The new corps was short-lived and did not play a significant role during the following Parthian campaigns due to the failure of Caracalla's anachronistic tactical experiments.

Prior to his death, during AD212–217, Caracalla had to face a new military threat on the northern borders of the Empire: the Alamanni. The latter were a confederation of Germanic tribes, and the name meant 'all men'. The Alamanni attacked the Roman frontier in its most exposed area: the *Agri Decumates* (Decumatian Fields) located in present day southwestern Germany, and linked up the limes of the Rhine with that of the Danube. The Decumatian Fields were the only place in continental Europe where the border of the Roman Empire was not marked by the Rhine or by the Danube. As a result, they were particularly exposed to foreign invasions and had been heavily fortified by the Romans. By conquering the Decumatian Fields an enemy of Rome could easily outflank the defences of the Rhine on the southeast or those of the Danube on the northwest.

The Alamanni settled just outside the Roman fortifications of the Decumatian Fields and started to attack them during the reign of Caracalla. Caracalla was able to contain the military activities of the newcomers by building new military infrastructure (such as roads) in the Decumatian Fields. This, however, had very significant financial costs for the Roman state.

During AD215–217, the Emperor campaigned with significant military forces in the Middle East, devastating a large portion of the Parthian Empire, much of it due to the internal divisions of those people. In AD217, however, Caracalla was killed during a plot organised by the commander of the Praetorian Guard.

The sudden death of Caracalla had extremely negative consequences for the military situation of the Empire, both on the northern borders – where the Alamanni could reorganise themselves in view of further incursions – and in the Middle East. Here the leader of the Praetorians who had organised the assassination of Caracalla, Macrinus, was proclaimed emperor but had to continue the ongoing war against the Parthians. The conflict came to an end only after the bloody Battle of Nisibis, which took place during AD217 and saw the Romans being defeated after three days of harsh fighting. Macrinus was greatly weakened by this significant military setback. Very soon, in fact, he had to face the revolt of Elagabalus (AD204–222), who was a cousin of the murdered Caracalla.

Elagabalus, thanks to the support of the Senate, dethroned Macrinus and rose to power. The reign of the new monarch, however, was brief since in AD222 he was assassinated and replaced with his young nephew Severus Alexander (d.AD235). The Battle of Nisibis was the last clash ever fought between Parthians and Romans. Soon after it, the Parthian Empire crumbled and the new Persian dynasty of the Sassanids rose to power in present day Iran.

The Sassanids had the ambition of expelling the Romans from the Middle East and thus the new Emperor Severus Alexander was soon forced to face them on the field of battle. During AD231–232 the

Roman Army campaigned against the Sassanids, obtaining some successes but they were by no means decisive. Severus Alexander's war against the renewed Persian Empire showed the Roman leaders that the Sassanids were much more powerful and dangerous than their Parthian predecessors. In AD235, the Emperor had to face a massive invasion of the Alamanni on the northern limes, but before he could face the enemy he was killed by his own soldiers who no longer considered him a capable military leader. The assassination of Severus Alexander, as we have already seen, marked the beginning of the terrible 'Crisis of the Third Century'.

Above left: A Roman auxiliary slinger from the Balearic Islands. (Marc Seriol (@marcmarkhus_photo), Legio II Traiana Fortis – Cohors I Barcinonum, Barcino Oriens)

Above right: A Roman auxiliary cavalryman armed with spear and javelins. (David Burns, Legio XIII Gemina)

Chapter 6

The Crisis of the Third Century, AD235–284

Conventionally the assassination of Emperor Severus Alexander is considered as the starting point of the crisis, but the internal and external troubles of the Roman Empire had started long before AD235. The first signs of military difficulties, for example, had been apparent during the reign of Marcus Aurelius when the Marcomannic Wars (AD166–180) had shown how the Germanic tribes of northern and central Europe were becoming a serious menace to the stability of the Roman frontiers. Since the reign of Augustus, the Romans had abandoned their ambitions to occupy modern day Germany, but had been able to contain the Germanic peoples and to counter their incursions in an effective way. The border between the Roman Empire and the barbarians had remained quiet and stable for a long time; the military supremacy of the Romans had never been contested and the relations with the Germani had generally remained quite positive.

Mass migration

During the second half of the second century AD, however, this situation changed in a dramatic way. Coming under pressure from war-like peoples migrating from the steppes of central Asia and eastern Europe, the Germani started to migrate towards the borders of the Empire. Now, for the first time in their recent history, the Romans had to face massive migrations. The defensive system, based on garrisons of legions and locally recruited auxiliaries, had no hope of countering the massive number of Germanic people now on the move. The border was too vast to be defended in an effective way, since simultaneous attacks from several different tribes became quite common. The Germanic peoples were effectively a confederation of tribes, and could assemble large numbers of warriors for attacks in a precise sector of the Roman limes. During the Marcomannic Wars, the Romans experienced new military difficulties: though ultimately, they were able to defeat the Germani, but with incredible difficulty and after suffering severe human losses. Added to this, the Antonine Plague killed thousands across the Empire.

The rise of the Sassanids

Another great political change, happening in the eastern part of the Empire, was to have very important consequences for the military situation of the Romans. In AD228, the Parthians, who had controlled present day Iran since the fall of the Seleucid Empire, were decisively defeated by a new Persian dynasty that assumed control of their territories. The new rulers of Iran were the Sassanids, who soon transformed the former Parthian Empire in a decisive way. The Parthians, coming from the steppes of central Asia, had created a feudal state that was a confederation of several different regions having large autonomy. The Sassanids, instead, reorganised the empire as a centralised nation with a strong ruler and a single religion. In doing this, they presented themselves as the heirs of the ancient Achaemenids, who had created a solid and immense multi-national empire. As a result of all these changes, the eastern part of the Roman Empire started to be menaced by a new threat that was much more dangerous than the Parthians. The Sassanids had a political organisation that was, in many aspects, very similar to that of the Romans and could easily confront the Roman Empire on almost equal terms. If the Parthians had

A Roman legionary cavalryman with hexagonal shield. (Legio XI C.P.F. Hispaniensis)

generally limited themselves to raids and guerrilla campaigns, the Sassanids were capable of launching full-scale invasions of the rich Roman eastern provinces. Since the ascendancy of the new dynasty, the state of war on the eastern limes was almost continuous.

Military anarchy and social change

With the ascendancy of the new Severan Dynasty, the same institutional nature of the Roman Empire was changed. Beginning with Septimius Severus, the Roman emperors began to be military despots, who were able to retain power due to the decisive support of the army and of the equestrian social class. The traditional respect that emperors had always showed towards the Senate started to disappear.

The monarch had absolute power and could easily impose his will by the use of violence. The senatorial class gradually lost importance, especially among the military. Loyalty towards the supreme leader became increasingly more important than real virtue. Since the soldiers were the effective holders of power, emperors started to treat the army as their own private property. Large donations of money were distributed to the legionaries and their pay was gradually augmented, and the number of troops under direct control of the emperor was increased.

The entire period of the Crisis of the Third Century is also known as the Military Anarchy, because it saw total chaos in the successions of emperors caused by the increasing power of the military. Since the reign of Septimius Severus, the Roman monarchs were chosen by the Praetorian Guard or by the army. The dynastic principle, which had worked quite well until then, was abandoned after the assassination of Severus Alexander (who was killed by his own troops). The legions, which were now mostly commanded by officers from the equites social class, began to kill and proclaim emperors according to their own interests. It was enough to offer a large donation to the Praetorians or to the legionaries to obtain the throne, while it was enough to reduce the pay of the soldiers to be deposed or killed. The weakness of the senatorial class and the continuous state of military emergency gave all political powers to the military leaders, something that temporarily ended only with the ascendancy of Diocletian in ad284.

The Crisis of the Third Century was not only a political and military one: first of all, it seriously reduced the demographic capabilities of the Empire due to the human losses caused by terrible epidemics. Economy and commerce were severely damaged, while the traditional Roman way of life had to change in a notable way. New religions became gradually dominant and Roman society completely changed its

Above left: A Roman legionary cavalryman with bronze helmet. (Legio XI C.P.F. Hispaniensis)

Above right: A Roman legionary cavalryman with oval shield. (Legio XI C.P.F. Hispaniensis)

internal organisation. Traditionally, since the late Republic, the Romans had developed a social structure that was based on three main classes: aristocracy (senatorial families), equites (the middle class) and the plebs (the lowest social group). The enormous difficulties of the crisis reduced the number of social classes from three to two: the richest members of society, the aristocracy and the wealthiest equites, started to form a new group known as *honestiores*. The modest equites and the plebs, instead, started to be known as *humiliores*. In practice, the crisis had cancelled the middle class: the inhabitants of the Empire were now extremely rich or extremely poor. The most important equites had transformed themselves into a new ruling class, gradually making the traditional aristocracy of the senators more marginal. This had very serious effects on the leadership of the army.

However, before this crisis began, another important social change had taken place. In 212, the Constitutio Antoniana of Emperor Caracalla gave Roman citizenship to all the inhabitants of the Empire. Until that moment, the Roman military organisation had been based on the distinction between citizens and non-citizens. With the promulgation of the Constitutio Antoniana, the main social difference existing between legionaries and auxiliaries was cancelled. A provincial man now did not have to serve in the army as an auxiliary for years in order to obtain Roman citizenship.

Finally, it is important to note how the crisis had a deep impact over the urban centres of the Empire. Many historians have defined Roman civilisation as an 'empire of the cities', because urban centres were the core of the Roman state. The crisis changed all this: the large, open and rich cities of the previous decades were progressively abandoned for fear of foreign attacks; new smaller cities started to be built, which were protected by defensive walls and thus were much more isolated and poor than before. Rome was encircled by new walls, with construction beginning under Emperor Aurelian. The western provinces became increasingly poor due to being exposed to foreign attacks. The eastern provinces became increasingly wealthy thanks to their superior agricultural capabilities. The economic decline of the western territories affected the monetary system, which started to have a decreasing circulation of gold and silver coinage in the western provinces. This altered the commercial balance of the Roman Empire's market in a decisive way, progressively destroying the economic integration of the Roman territories.

A period of chaos

In AD235, the leadership of the Roman Empire was taken by Maximinus Thrax (d.AD235), who commanded one of the legions fighting against the Alamanni when Severus Alexander was killed. The new monarch, being an experienced military commander, obtained a series of significant victories over the Alamanni in the Decumatian Fields. As soon as the situation on the Rhine had been stabilised, however, he had to move on the Danube to counter the incursions of the Sarmatians.

During AD238, a major internal revolt broke out in Roman Africa, where a local governor named Gordian was proclaimed emperor by his troops; Maximinus Thrax was able to crush the rebellion quite rapidly, but completely lost the support of the Senate. In AD238, the senators revolted against his rule and proclaimed the grandson of Gordian – Gordian III (AD225–244) – emperor. Maximinus Thrax then tried to invade the Italian Peninsula from the north, by moving from his strongholds in Pannonia; while besieging the strategic city of Aquileia, however, he was killed by his own soldiers. Gordian III ruled for less than two years and was not able to face the new threats that were emerging on the borders of the Empire. He had to fight a new war against the Sassanids, who were guided by the ambitious Shapur I (AD215–270). Initially he obtained some significant victories, for example at the Battle of Resaena, but in AD244 he was defeated at the Battle of Misiche and died during the latter clash.

Gordian III's successor was the commander of the Praetorians, Philip the Arab (d.AD249). The latter soon made peace with Shapur I and had to face the Germanic tribe of the Carpi on the Danube. In AD249, after the hostilities with the Sassanids had resumed and after the Roman defences on the Danube had been broken up in various points, the legions of Pannonia acclaimed Decius – who had been one of Philip the Arab's main collaborators – as the new emperor. The troops still loyal to Philip were defeated in northern Italy and thus Decius assumed power in Rome.

Decius was the first Roman monarch to face the Goths on the Balkan frontier, but could do very little to stop the raids and incursions of the newcomers. In AD251, at the Battle of Abritus, a gigantic clash took place between the Roman Army and the Goths. The latter, against all odds, were able to prevail despite suffering severe human losses; Decius, after having fought with courage, was killed on the field of battle: he was the first Roman emperor to die during a battle fought against a foreign enemy. Decius's successor was Trebonianus Gallus (b.AD206), who soon made peace with the Goths in order to stabilise the Danubian frontier. During his two years of reign, the new monarch saw two consecutive Sassanid invasions of his eastern provinces and several barbaric incursions across the Danube.

The Roman troops, dissatisfied with the leadership of Trebonianus Gallus, rose up in revolt; the ensuing civil war ended only when the Emperor was betrayed and assassinated by his own supporters. Following these events, the Empire entered a period of complete political chaos: Aemilian (AD207–253), an important commander of the Roman troops stationed in the northern Balkans, ruled as emperor for just two months before being killed by his own soldiers who favoured the ascendancy of another general – Valerian – to the throne. The latter ruled until AD260 and spent most of his reign fighting against the Sassanids of Shapur I, who had occupied the Roman

Above left: A Roman legionary cavalryman with parade mask. (Legio XIII Gemina)

Above right: A Roman auxiliary cavalryman with parade helmet and shield. (Robert Vermaat/Ala I Batavorum)

Middle East. Valerian could count on the support of his son, Gallienus, who was a very capable military commander; despite this, however, he was soundly defeated by the Sassanids at the Battle of Edessa in 260. The latter was one of the worst military disasters ever suffered by the Romans, since it saw the destruction of a large part of the Roman Army and the capture of Valerian. Valerian, after having being humiliated by Shapur I, died, probably in AD264 as a captive of the Sassanids. The defeat and capture of Valerian led to the beginning of the worst and most chaotic phase of the Crisis of the Third Century.

Valerian's son and successor Gallienus, had to face the outbreak of a series of internal revolts. In the eastern provinces, which were no longer protected by the Roman troops and thus were now exposed to Sassanid conquest, the Roman client king of Palmyra – Odaenathus (AD220–267) – assumed control over all the lands that had previously been under imperial control. At the same time, in the north, the overall commander of the troops stationed on the Rhine frontier – Postumus (d.AD269) – rebelled against Gallienus. Postumus soon assumed control over most of the Empire – Gaul, Britannia and Iberia – and then seceded from

A Roman auxiliary cavalryman with scale armour. (Robert Vermaat/Ala I Batavorum)

the Roman state with his own territories. He did not make any effort to depose Gallienus in Rome. Instead, he created an autonomous state known as the Gallic Empire that had its own parallel institutions modelled on those of the Roman Empire's central government.

The Gallic Empire had its own army, Praetorian Guard, coins, consuls and senate. Both Postumus and Odaenathus proved to be very capable military leaders. The first defeated in a decisive way both the Alamanni and the Franks, restoring order in the Rhineland; the second, instead, defeated Shapur I and freed the Roman Middle East (Syria and Mesopotamia) from Sassanid rule. In AD262, he even besieged the Sassanid capital of Ctesiphon before being proclaimed 'King of Kings of the East'. During AD267, Odaenathus was assassinated for unknown reasons; he was succeeded by his infant son Vaballathus (AD259–74), who reigned under the regency of his mother Zenobia (AD240–274).

Postumus, was killed during a rebellion of his own troops in AD269. While these events took place in the provinces that had seceded from the Empire, Gallienus struggled to defend the Italian Peninsula and the Balkans from the destructive incursions of the Germanic peoples. Thanks to his intelligent reform of the military forces, he was able to obtain some significant victories over the foreign invaders. In 268, however, his cavalry commander Aureolus (AD220–268) revolted against him. A new civil war broke out, which ended with the defeat of Aureolus but also with the assassination of Gallienus. The latter was succeeded by Claudius Gothicus (AD214–270), an experienced military commander who obtained a series of great victories. In AD268, the new monarch defeated the Goths at the Battle of Naissus and then the Alamanni at the Battle of Lake Benacus. In AD269, Claudius Gothicus reconquered Iberia from the Gallic Empire; during the early months of AD270, however, he died after having appointed his cavalry commander – Aurelian (AD214–275) – as his successor.

The victories of Aurelian

Aurelian was without doubt one of the greatest Roman monarchs. In just five years, he was able to re-unify the Empire and temporarily stabilise its borders. First, he defeated both the Alamanni and the Goths in order to secure his position in Italy and in the Balkans. He then decided to evacuate the province of Dacia, which was too exposed to foreign invasions due to the fact that it was located north of the Danube. In AD272, Aurelian turned on the newly formed Palmyrene Empire ruled by Zenobia, which had come to comprise most of the Middle East: southeastern Anatolia, Syria, Palestine and even Egypt. Aurelian knew that his state could not survive without the massive amounts of grain produced in Egypt and thus organised a massive offensive against Palmyra. Moving across Anatolia, the Roman troops reconquered one province after the other, obtaining some spectacular victories. Then they besieged the same city

A Roman auxiliary cavalryman wearing parade helmet with facial mask.
(Robert Vermaat/Ala I Batavorum)

of Palmyra, which was captured together with Zenobia. Palmyra was razed to the ground, while Zenobia – in chains – participated in Aurelian's triumph in Rome.

In AD274, the Emperor turned his attention to the Gallic Empire, with the ambition of destroying it. He assembled a large army and invaded Gaul, with the objective of facing his enemies in a large pitched clash. The latter, known as Battle of Chalons, took place in northern France during the early months of 274. Tetricus, the new leader of the Gallic Empire, was severely defeated and forced to surrender. Soon after these events, the Gallic Empire dissolved and its territories were re-annexed by the Roman Empire.

In AD275, while preparing a campaign against the Sassanid Empire, Aurelianus was assassinated by one of his collaborators for futile personal reasons. The unexpected death of Aurelianus opened a new phase of political chaos, which saw the election of two emperors – Tacitus and Florianus, two half-brothers – who reigned for just a few months. In June AD276, Probus, who had been one of Aurelianus's most important military commanders, became emperor with the support of the Senate. He was a very capable military leader, as shown by the many victories that he achieved over the Alamanni, Franks, Burgundians and Goths. Probus campaigned on the territories of his Germanic enemies and repaired the Roman defensive structures in the Decumatian Fields. In 282, however, he was assassinated as part of a military coup. Probus's successor was Carus, who ruled for just a few months during which he was able to obtain some victories over both the Sarmatians and the Sassanids. Carus probably died of natural causes and was succeeded by his son Carinus. The latter, however, soon had to face the ambitious Diocletian who was proclaimed emperor by the Roman troops of the eastern provinces in 284 and defeated Carinus during 285.

Chapter 7

The Roman Army of the Third Century

Gallienus, who ruled the Roman Empire during AD260–268, is commonly considered as the first Roman monarch who understood that the military organisation created by Augustus had to be changed in order to have a more effective defence of the Empire. The traditional structure consisting of legions and auxilia units was too static to counter the massive migrations of Germanic peoples or the large-scale invasions of the Sassanids. In addition, civil unrest and internal threats had

A Roman legionary cavalryman with parade helmet and armour. Spears and javelins with painted shafts were used only on parade or during hippika gymnasia, or cavalry tournaments. (Legio XI C.P.F. Hispaniensis)

to be taken into consideration. Until the beginning of the Crisis of the Third Century, the Roman Army had only been employed to fight on the frontiers, but the experiences of the internal secessions had clearly shown the need to use the armed forces to perform police functions. One of the consequences of the crisis had been the birth of a new phenomenon of internal unrest, a sort of brigandage that saw the formation of large bands of insurgents/criminals (commonly known as *Bagaudes*). As a result, the army had to be ready to maintain public order in addition to serving on the frontiers.

The Constitutio Antoniana, finally cancelled the auxilia system. Gallienus tried to face and resolve all these problems, starting with the defence of the frontiers. The idea of a static line of defence had been totally surpassed by events. The Empire needed central reserves of highly mobile troops, who could move very rapidly and counter different menaces in several points of the frontiers. The only units that could deploy such needed levels of mobility were the cavalry: the legionary or auxiliary infantry garrisoned on the borders had no chances of fighting with success against simultaneous threats. Initially, Gallienus created four small cavalry reserves, located in strategic places of the Empire: Mediolanum in northern Italy, Sirmium in Serbia, Poetovio in Slovenia and Lychnidos in Macedonia. Over time, the two centres of Mediolanum and Sirmium became the most important ones, where two large strategic reserves of cavalry were stationed. The troops in Mediolanum were ready to counter any emergency menacing Italy; the troops in Sirmium were ready to act on the long Danubian limes.

Military reform

As a result of Gallienus's cavalry reform, mounted units became much more important. As we have already seen, the Romans had never considered cavalry as a very important part of their military forces: the Constitutio Antoniana, however, had transformed all provincials into Roman citizens and thus new cavalry units could now be easily recruited in all the provinces of the Empire. The collapse of the auxilia system and the contemporary expansion of the cavalry led to the birth of several new categories of mounted troops, which made up the new strategic reserves and which had some innovative characteristics.

Before analysing these new troop types, however, it is important to consider the other military reforms produced by Gallienus. These mostly affected the leadership of the army: the senatorial class was formally excluded from any form of military service. The aristocracy, which for centuries had controlled the entire Roman military machine, was in such a state of decay that the Emperor could assign command of the whole military apparatus to the richest members of the equites. In effect, Gallienus was not revolutionary, but just transformed an informal practice into a formal one. Since the time of Marcus Aurelius, the senators had gradually and voluntarily abandoned the role of legions' commanders, preferring to dedicate themselves to much more lucrative activities such as trading. The traditional ethos of the Roman aristocrats had disappeared together with the ancestral values of Roman society. Luckily for Rome, however, this important heritage was accepted by the equestrians of the new honestiores social class, who were ready to take the place of the senators in the ranks of the army. The Legatus Legionis was now substituted with the *Praefectus Legionis*, who was a member of the equites. The general expansion of cavalry was not limited to the creation of new separate mounted units, but also involved the transformation of legionary cavalry. Since the time of Augustus, the cavalry of each single legion consisted of 120 horsemen divided into four turmae with 30 soldiers each. With the reform brought in by Gallienus, the cavalry of each legion was expanded to the establishment of an *ala miliaria* (24 turmae of 30 men each, for a total of 720 horsemen). The new types of cavalry units created by Gallienus were the direct heirs of the former mounted auxilia. Very little is known about their internal organisation, but what we know is that they all assumed the general denomination of 'vexillationes'. Since the reign of Gallienus, in fact, this term was used to identify detachments taken from the legions; the 'vexillatio' was now the standard cavalry unit, which could be formed by different kinds of equites.

A Roman auxiliary cavalryman. Note the decorative parade harness worn by the horse. (Robert Vermaat/Ala I Batavorum)

Equites Dalmatae: were the new cavalry vexillationes created by Gallienus. These originated as detachments of existing mounted units (both legionary cavalry and auxilia) that were separated from their parent corps in a definitive way and thus constituted new independent units. It is highly probable that this process was caused by the military emergency experienced by the Roman Empire during the reign of Gallienus. With the division of the Empire caused by the two large secessions of Gaul and Palmyra, only the central territories of Italy and the Balkans remained under direct control of Gallienus. As a result, many detachments coming from Gaul or the eastern part of the Empire that were temporarily stationed in the Balkans found themselves isolated from their parent units. Gallienus transformed these into independent units and thus the term vexillationes gradually started to indicate independent mounted units. The Equites Dalmatae were probably the first new category of cavalrymen to be formed, since they were quite numerous. The *Notitia Dignitatum*, an important primary source produced several decades after Gallienus's cavalry reform, lists 48 units of this kind. The term 'Dalmatae' had no specific ethnic meaning, because it was simply used to indicate the geographical provenience of these units: they were originally stationed in the region of Dalmatia, but were composed of soldiers coming from every corner of the Empire. It would be wrong to think that these units were entirely raised from inhabitants of the Balkan provinces, because this would not explain the survival of this kind of unit after the end of the Third Century Crisis's military emergency. It is true that the Balkan provinces were the real military core of Gallienus's truncated Empire, but they could not deploy such a large number of cavalry units alone. With time, the original detachments stationed in the Balkan provinces under Gallienus started to be located in other areas of the Empire. However, they retained their original denomination of Dalmatae to show that they had formerly been part of Gallienus's forces. During the terrible years of the secessions, the Emperor had to face a severe military crisis and was obliged to use all the military resources available in the territories under his control. The Equites Dalmatae were permanently assigned to the garrisons of frontier provinces, as highly mobile cavalry forces that could counter enemy incursions.

Equites Illyriciani: these had the same origins as the Equites Dalmatae, but were initially stationed in a different geographical area. While the Equites Dalmatae were detachments garrisoned in Dalmatia, the Equites Illyriciani came from Illyricum. Dalmatia and Illyricum were the heart of the Danubian provinces in the Balkans, where many vexillationes who came from other areas of the Empire were located. When Gallienus had to build up a new army from his truncated territories, he used cavalry detachments as the backbone of his new military forces and transformed them into independent units. The *Notitia Dignitatum* lists 23 units of Equites Illyriciani, garrisoned on various border provinces of the Empire; like the Equites Dalmatae, they started to be used as frontier cavalry during the reign of Diocletian.

Equites stablesiani: their history and development was exactly the same as the Equites Dalmatae and Equites Illyriciani, the only difference being the regional provenence of the units. While the previous two categories of troops were formed by converting detachments that were stationed in Dalmatia or Illyricum, the Equites stablesiani were created by converting the vexillationes garrisoned in northern Italy into independent cavalry units. As we have already seen, Italy and the Balkan provinces were the only territories that remained under direct control of Gallienus. If we look at the original four strategic cavalry reserves created by the Emperor, we can clearly see that they were located in northern Italy and the Balkans: Mediolanum in northern Italy, Sirmium in Serbia, Poetovio in Slovenia and Lychnidos in Macedonia. It is thus reasonable to suppose that the Equites stablesiani were based in Mediolanum, the Equites Dalmatae in Sirmium and Poetovio, the Equites Illyriciani in Lychnidos.

The term stablesiani comes from the particular title of 'stabulensis' that was given by Gallienus to Aureolus, who was the first commander of the strategic cavalry reserve based in Mediolanum. Stabulensis has an honorific meaning, because it designated Aureolus as the senior officer in charge of the imperial stables. In total, the *Notitia Dignitatum* lists 16 units of Equites stablesiani. Like the similar units of Dalmatae and Illyriciani, the Equites stablesiani later became frontier cavalry.

***Equites promoti*:** as we have already seen, Gallienus augmented the numerical composition of the legionary cavalry to the establishment of an ala miliaria (24 turmae of 30 men each, for a total of 720 horsemen). As a result, the cavalry of each legion now had the necessary numbers to act in a more independent way. During the military emergency caused by the invasions of the Germanic tribes and by the internal secessions, Gallienus transformed several units of legionary cavalry into independent ones: that is the reason why these were known as promoted. It is reasonable to suppose that the Equites promoti retained their original internal organisation, at least initially. The *Notitia Dignitatum* lists 31 units of Equites promoti. Like the other new cavalry units created by Gallienus, the Equites promoti were progressively garrisoned in all the provinces of the Empire and used as frontier cavalry.

***Equites indigenae*:** during the early centuries of the Empire, the great majority of the Roman Army's cavalry had been made up of provincial auxiliaries. When the Constitutio Antoniana gave Roman citizenship to all provincials, the status of these auxiliary cavalrymen suddenly changed. Most of the cavalry units formed by raising local elements in the various provinces started to be defined as *indigenae* (indigenous), abandoning their auxiliary status but retaining their local nature. In total, the *Notitia Dignitatum* lists 38 units of native cavalry, stationed in every corner of the Empire (mostly on the frontiers). Every province had its own units of native cavalrymen, who were strongly influenced by the local military traditions of their territories.

***Equites Mauri, Equites scutarii, Equites sagittarii*:** these three categories of cavalrymen were not directly related to the reforms of Gallienus, but they were part of the new cavalry organisation that emerged after the changes of the late third century. Moorish (ie, Numidians and Mauri) light horsemen had always been a fundamental component of the Roman Army's auxiliary cavalry. Their role was particularly important during the campaigns fought against the Parthians or the Sassanids in the east, because they were excellent skirmishers armed with javelins and mounted on fast horses. The Moorish light skirmishers were the perfect Roman tactical response to the large numbers of horse archers deployed by the Parthians and the Sassanids. They remained a military elite after the Crisis of the Third Century. As a result, although the auxilia system had disappeared, we find ten Moorish light cavalry units listed in the *Notitia Dignitatum*. These were garrisoned in different areas of the Empire and formed a distinct category of troops (separated from that of the Equites indigenae even in the provinces of North Africa, which were the home of the Moorish tribes). The existence of a specific denomination for Moorish units confirms the high consideration that the Romans had of them, especially during a period that was characterised by a strong need for new light units (both mounted and on foot). The *Notitia Dignitatum* includes eight units of Moorish infantry, a clear signal of the fact that the Mauri fighters were highly appreciated as foot skirmishers.

The Equites scutarii are a category of mounted troops about which we know very little. The term 'scutarii' does not mean that they were all equipped with shields; during the last centuries of the Empire, it was used to indicate guard units. It is thus plausible to presume that the 17 units of Equites scutarii listed in the *Notitia Dignitatum* acted as mounted bodyguards, although we have no clear idea of their effective functions. Maybe they were used as the mounted escort of high-ranking officers.

Finally, some words on the Equites sagittarii; the difficult wars fought against the Sassanids convinced the Romans to include some horse archers in their army. These new units were formed in all the provinces of the Empire and the *Notitia Dignitatum* lists 51 of them.

The Roman military reforms of the third century were not limited to the cavalry; the legionary infantry was deeply affected by them, especially light troops. As we have already seen, the Roman Army had employed provincial soldiers in order to form the majority of its light infantry units; it should not be forgotten, however, that the legions had always comprised a certain number of lightly equipped infantrymen. These were known as *antesignani* and there is evidence of them since the times of Julius Caesar. Antesignani are elite light infantrymen who made up the vanguard of each legion; they were frequently used for special missions and carried lighter equipment than the standard heavy infantry legionaries. They were not part of the heavy infantry formations and used small oval shields instead of the larger ones carried by standard legionaries. During marches, their main functions were

Above left: A detail of the decorated chanfron worn by Roman military horses during parades and tournaments. (Robert Vermaat/Ala I Batavorum)

Above right: A Roman auxiliary cavalryman with oval shield. (Robert Vermaat/Ala I Batavorum)

those of exploring the territory and of protecting the flanks of the legion. In battle, they countered the missile troops of the enemy and supported cavalry during its attacks thanks to their superior mobility.

It is important to note that despite their important functions, the antesignani did not exist as a separate corps inside the legions. They were not a standing unit but were drawn from the ranks of the cohorts when tactical considerations demanded it. Their assignments, albeit very important, were always temporary.

The antesignani were usually drawn from the youngest and fittest legionaries of each cohort, in order to perform special missions that needed a high level of mobility.

During the third century AD, particularly under the reign of Septimius Severus, the light infantry component of the legions started to be increasingly important and acquired a stable organisation. The first legions to have a fixed number of light infantrymen, who performed as skirmishers on a permanent basis, were the three new ones created by Septimius Severus for his Parthian campaign. Since their first encounters with the Parthian mounted archers, the Romans had clearly suffered from a lack of missile troops inside their heavy infantry formations. Septimius Severus tried to resolve this problem in a definitive way by including a fixed number of light infantrymen in his three newly raised elite legions. According to the new structure introduced for the latter and later adopted by all the legions, each cohort was to include two centuries of light infantry and four of heavy infantry (the elite Cohors Prima had two centuries of light infantry and three of heavy infantry). The transformation of two centuries from heavy to light infantry clearly shows the increasing need for missile troops. The new Roman light infantrymen were known as *lanciarii*, from the name of their main weapon (the 'lancea'). Quite frequently, on special occasions, all the centuries of lanciarii included in a legion could be brigaded together and detached in order to act independently. This means that, like the former antesignani, they were also considered elite troops. Over time, however, the lanciarii gradually transformed themselves into elite guard units and partly abandoned their original nature of light infantry soldiers. By the time of Diocletian, they had completed this transformation and started to be grouped into large independent units. By the time of the *Notitia Dignitatum*, the Roman Army included nine legions entirely formed by lanciarii.

During the turbulent decades of the third century, the Roman generals started to make increasing use of irregular units entirely composed of barbarian soldiers. Germanic warriors were generally very happy to serve in the Roman

A Roman legionary cavalryman wearing the subarmalis, the protective padded garment worn under the metal armour. (Legio XI C.P.F. Hispaniensis)

Army as mercenaries, but were always quite reluctant to integrate themselves into regular military units. As a result, the Romans created two new kinds of military units that were entirely composed of barbarian fighters: the infantry *numerus* and the cavalry *cuneus*. During the following centuries, these kinds of units were opened to Romans and started to receive a more formal internal organisation. During the third and early fourth centuries, they remained corps of barbarians with a sort of semi-regular nature. Both the numerus and the cuneus had no standard structure to speak of; the groups of German warriors could be very numerous or few and usually had no internal subdivisions. What we know for sure is that the members of numeri and cunei fought in their own style and often under their own chieftains (who contracted their services to the Roman military authorities). Initially, these units of hired barbarians were mostly employed along the borders, as a supplement to the units of auxilia guarding the frontier. Over time, they became increasingly numerous and important, transforming themselves into a significant portion of the Roman military forces.

The personal equipment of a Roman legionary cavalryman. (Legio XIII Gemina)

Chapter 8

Roman Military Equipment and Tactics

At the beginning of the Early Empire, the infantry of the Roman Army already employed a highly standardised panoply that comprised some basic elements: helmet, armour, shield, heavy javelin, short sword and dagger. This had been introduced with the military reform of Gaius Marius, and was implemented shortly before the outbreak of the Roman Civil Wars. Under Augustus, the panoply of the Roman heavy infantrymen was altered in some of its components, but did not change in a radical way. New models of helmet were introduced, the previous *lorica hamata* armour was replaced with the new *lorica segmentata* and the original oval *scutum* was substituted with a new rectangular one.

The tactics of the heavy infantry, however, remained more or less the same since the offensive weapons did not change. The new panoply continued to be used by the legionaries until the last years of the second century. The cavalrymen of the legions were equipped similarly to their foot companions, but continued to use the lorica hamata and oval shields although the latter were no longer employed by the infantrymen. The offensive weapons of the horsemen, however, consisted of spear and long sword instead of the heavy javelin and short sword used by the legionaries. Until the military reform of Augustus, the cavalry and light infantry of the auxiliaries did not have standard equipment since each group was usually equipped in its own 'national' way. With the development of the auxilia system, however, the many units of non-Roman auxiliaries that were created across the Empire received a standardised set of

A display of eight different types of *gladius*. (Legio VI Victrix Cohors II Cimbria)

equipment that comprised the following elements: helmet, lorica hamata armour, oval shield, spear for the cavalrymen or light javelins for the infantrymen and long sword for the cavalrymen or short sword for the infantrymen.

Helmets

At the beginning of the Imperial period, the most common model of Roman *galea* (helmet) was the Montefortino one. This was very easy to produce and the Romans had copied it from the Celts, introducing some minor adjustments that were needed to transform the Montefortino into the standard helmet of the legions. The name of this model, galea, derives from the Italian place where a first helmet of this kind was discovered. The Montefortino galea had a round shape with a raised central knob and a protruding neck guard; the former was usually surmounted by coloured plumes, while the latter was generally decorated with incisions. In addition, it had a pair of cheek pieces: these could be of two main different kinds, according to their shape. The first model of cheek pieces reproduced the stylised shape of the cheek; the second one was *trilobate*, consisting of three little discs having the shape of a triangle. Apparently, trilobate cheek pieces were influenced by the helmets and armour of the Italic peoples, which frequently included this peculiar combination of three discs. Over time, this kind of cheek piece was abandoned and the other one became dominant (being the one adopted by the Romans).

In general terms, Montefortino helmets had many positive features: they were easy to produce, they gave excellent protection to the wearer's face, they could be easily decorated in many different ways and they were comfortable enough to be worn for long time. All these characteristics made the Montefortino the most popular helmet of the Roman Republican period.

A display of five different types of gladius for an officer. (Legio XI C.P.F. Hispaniensis)

During the early decades of the Empire, another helmet designed by the Celts and later adopted by the Romans co-existed with the Montefortino: the so-called Coolus helmet. This had a simple round shape, with a ridge running around its base. On the back, this ridge enlarged to become a sort of neck guard, similar to the Montefortino. In addition, the Coolus helmet had a couple of cheek pieces that mirrored the human cheek in a stylised way. Being extremely easy and cheap to produce, the Coolus helmet became popular especially with the auxiliaries of the Roman Army. According to archaeological finds, cheek pieces were added to Coolus helmets at a later date, as a result of the Montefortino's influence. As a result, at least initially, the Coolus helmet had looked like a simple metal cap. Crest fittings were added to the helmet at a later time.

From the last years of Augustus's long reign, both the Montefortino and the Coolus were eventually replaced by the Imperial helmet. This was produced in two main sub-types – Imperial Gallic and Imperial Italic – which could in turn be divided into further sub-categories corresponding to different versions. The Imperial Gallic helmets, featured a pair of distinctive embossed eyebrows on the forehead region and were both carefully made and elaborately ornamented. They were produced by the skilled Celtic craftsmen of Gaul. The Imperial Italic helmets, lacking the eyebrows and being more roughly made – were produced by less-skilled copycats working in Italy. Over time, however, the differences between these two main sub-types started to disappear since the Imperial Italic helmets started to be of superior quality. Basically, the Imperial helmet was an improved version of the Montefortino one, having some more advanced features: sloped neck guard with ribbing at the nape, projecting ear guards, brass trim and decorative bosses. Following the combat experiences of Trajan's Dacian campaigns, two iron bars riveted crosswise were added across the helmet skull in order to reinforce it. As an alternative, two thick bronze strips might be riveted to the top of the helmet. From AD125, these modifications became permanent features.

The Imperial Gallic helmet was produced in three main versions: the original version 'G', the version 'H' that had a different style of eyebrows and a more sloping neck guard, the later version 'I' that had a 'twist-on' crest holder instead of the previous 'slide-on' one. The last version, differed from the previous two, and was made of brass and not of iron.

The Imperial Italic helmet was produced in four main versions: the original version 'D' and version 'E', which both had integral brass cross-braces placed flat against the skull but were decorated in different ways; the G had brass decorations applied between the crossbars; the H that was made of bronze instead of iron and had a deeper neck guard as well as embossed crossbars. This last version came into use around AD180 and had a dome-shaped knob where the cross-braces met at the crown of the head.

**The standard legionary gladius.
(Legio XIII Gemina)**

Chainmail

The Roman lorica hamata was an improved version of the Celtic chainmail, which was designed for the first time during the fifth century BC. This consisted of thousands of small iron rings linked together in order to form a mesh: the rings were interlocked creating something similar to a knitted sweater. Depending on its dimensions, it had an approximate weight of 10kg. Generally, Roman chainmails were sleeveless but had reinforcement panels for the shoulders: these were attached across the top of the back and held at the front by a bar-and-stud device. A double thong was stretched from the rings attached just above the inner corners of cut-outs to the outer corner of each of the reinforcement panels. The edges of the latter were bound with rawhide in order to create a raised border. The panels could have angled or rounded ends on the chest; they could be made of iron rings like the rest of the armour or could be of leather. The Roman lorica hamata comprised alternating rows of closed 'washer-like' rings punched from iron sheets and rows of riveted rings made from drawn wire that ran horizontally. These produced very flexible, reliable and strong armour. Each ring had an inside diameter of about 5mm and an outside diameter of about 7mm. Up to 30,000 rings would have gone into one lorica hamata and thus the estimated production time was two months, even with continual slave labour at the state factories. Although labour-intensive to manufacture this kind of armour, with good maintenance, could be used for several decades by a soldier. Constant friction kept the rings free of rust, unlike the lorica segmentata, which needed constant maintenance to prevent corrosion. Usually, the lorica hamata covered the legs to the knees. Until AD10–15, all the Roman soldiers wore the chainmail. During those years, however, the legionaries replaced it with the new lorica segmentata and thus the lorica hamata continued to be worn only by the cavalry and by the newly organised auxilia (both light infantrymen and cavalrymen).

The lorica segmentata (segmented cuirass) was no doubt the most advanced kind of armour produced in the Mediterranean during Antiquity. It consisted of metal strips fashioned into circular bands, which were fastened to internal leather straps. Thanks to archaeological finds, we know that the Roman legionaries fighting at the Battle of Teutoburg employed both the lorica hamata and the lorica segmentata. As a result, it's possible to say that the segmented armour became of common use during the last years of Augustus's reign. The strips of the lorica segmentata were made by overlapping ferrous plates that were riveted to straps made from leather. The plates were made of soft iron on the inside and of rolled mild steel on the outside, in order to be hardened against damage but without becoming brittle. They were hardened by packing organic matter tightly around them and by heating them in a forge, in order to transfer carbon from the burnt materials into the surface of the metal.

A legionary gladius with decorated bronze fittings on the scabbard. (Legio XIII Gemina)

The plates were obtained from beating out ingots. The strips of each cuirass were arranged horizontally on the body, overlapping downwards; they surrounded the torso in two halves, being fastened at the front and at the back. The upper body and the shoulders were protected by additional metal strips. Because of its form, a lorica segmentata was compact to store. It was possible to separate it into four sections, each of which would collapse on itself by forming a compact mass.

The fitments that closed the various plate sections together could be of many different kinds (buckles, lobate hinges, hinged straps, tie-hooks and tie-rings) but were all made of brass. These started to be replaced by simple rivets at a later date; in addition, the lowest plates of the cuirass were replaced by a single and broader plate. Since all the components of the lorica segmentata moved in synchronisation, this kind of armour was extremely flexible. It had, however, two points of weakness: first, a legionary needed the help of a comrade to put it on; second, it could be severely damaged by rust, so had to be kept clean.

During the Dacian campaigns, in order to protect their arms and legs from the *falx* (sickle) used by most of the enemy warriors, the Roman legionaries started to wear additional protective elements together with their lorica segmentata. For protection of the arms, they used the *manicae* (arm-guards) made up of curved and overlapping metal plates that were fastened to leather straps. To protect the legs, they used simple metal greaves. A single manica consisted of one shoulder plate, of about 35 strips made of iron or bronze, 90–120 leathering rivets, three or four strips of leather and one padded lining. The metal strips were longer at the top of the arm and all had holes at their lower edge, through which rivets passed from the inside to hold the straps in place. The lower plates of a manica, overlapping upwards like all the others, could sometimes be riveted together in order to be stronger.

Scale armour
In addition to the lorica hamata and lorica segmentata, the Roman Army used another two models of armour: the *lorica squamata* (scale armour) that was extremely popular in the eastern regions of the Empire and the muscle cuirass that was worn by the Praetorian Guard and by the officers.

The lorica squamata was made from small metal scales sewn to a fabric backing. The squamae (individual scales) could be of iron or bronze; the metal was generally not very thick, 0.5mm to 0.8mm being a common range. Since the scales overlapped in every direction, however, the multiple layers gave good protection. Scales could have rounded, pointed or flat bottoms with the corners clipped off at an angle. They could be flat, or slightly domed, or have a raised midrib/edge. The scales were wired or laced together in horizontal rows that were then laced or sewn to the backing. Therefore, each scale had from four to 12 holes: two or more at each side for wiring to the next scale in the row, one or two at the top for fastening to the backing, and sometimes one or two at the bottom to secure the scales to the backing or to each other. Sometimes the squamae could be tinned.

Scale armour was used by the legionaries but also by the auxiliaries coming from the eastern provinces, including light infantrymen and archers.

Cuirass
The muscle cuirass was a type of body armour cast to fit the wearer's torso and designed to mimic an idealised human body. It was cast in two pieces, the front and the back, which were later hammered together. The cuirasses of officers were highly ornamented with sculpted mythological scenes or with similar rich decorations. Usually, the muscle cuirass was used in combination with fringed stripes of leather known as *pteruges*, which were worn at the armholes and at the lower edge of the cuirass.

Left: **An officer's gladius with decorated silver fittings on the scabbard. (Legio XIII Gemina)**

Below: **Seven different types of pugio. (Legio VI Victrix Cohors II Cimbria)**

Shields

Until the adoption of the lorica segmentate, the Roman legionaries used oval shields, which were copied from the Celtic ones. These continued to be used by the cavalry and by the auxiliaries even after the legions adopted the new rectangular shields during the late years of Augustus's reign. The oblong Roman oval *scutum* (shield) originally had a central spine made of wood and an *umbo* (boss) made of metal, which was designed to reinforce the whole structure of the shield. Bosses were oblong and could have different shapes. The main body of the shield was made of oak planks, which were chamfered to a thinner section towards the rim. The wooden spine, swelling in the middle, was shaped in order to correspond with a round or oval cut-out in the shield's centre. The strap-type metal boss crossed over the wider section of the spine and was riveted on the external surface of the shield. The latter, on both sides or only on the front one, was entirely covered with leather that could be painted in various bright colours and have decorations of several kinds. Bosses corresponded to the handle of the shield on the back and thus had a fundamental function in protecting the user's hand. Additional metal binding was frequently attached to the external edges of the shield, in order to reinforce it.

The oval shields carried by the light infantry and cavalry were generally a bit smaller than the heavy infantry ones and had a more circular shape; those used by the horsemen, and in many cases, could also be hexagonal in shape and not oval. The rectangular scutum that replaced the previous oval one was obtained from three sheets of wood glued together and covered with canvas. Over the canvas there was an external surface made of leather. The rectangular shield was convex and had a spindle-shaped boss along its vertical length. With an average weight of 6kg it was 105.5cm high, 41cm wide and 30cm thick due to its semi-cylindrical nature.

The mass production of the scutum was something quite complex, considering the various phases that were needed to create a shield of good quality. Its features, however, gave a great tactical advantage to the legionary over possible opponents. The rectangular shield was light enough to be held in one hand and was long/large enough to cover the entire wielder (making him very unlikely to get hit by enemy

Above left: A standard legionary *pugio*. (Legio XIII Gemina)

Above right: A legionary's pugio with bronze scabbard. (Legio XIII Gemina)

missile weapons). In addition, the metal umbo in the centre of the shield made it a punching weapon that could be used during hand-to-hand fighting. During the Dacian campaigns, metal edges were added to all Roman shields, since the *falx* of the Dacian warriors could easily penetrate and rip through a scutum. Most of the Roman shields were painted red and were decorated with highly standardised devices including eagle wings, Jupiter's lightning bolt and laurel wreaths.

Javelins and spears

The first weapon to be used by a Roman legionary in battle was the *pilum* (javelin) specifically designed to kill enemies from long distance or to limit them in the use of their shields. The pilum, in fact, was extremely difficult to remove after hitting the external part of a shield or of a cuirass. It had a barbed head

and a long narrow socket or shank, made of iron mounted on a wooden haft. The barbs were designed to lodge in the enemy shields so that the whole javelin could not be removed. The long iron shank prevented the head from being cut from the shaft. The total weight of a pilum was between 1 and 2kg. As a result, this kind of weapon had a great power of penetration.

The shank, due to its peculiar function, was made of softer iron than the hard pyramidal tip. After hitting an enemy shield, it would bend in order to render the shield useless. The shank of the pilum was joined to the wooden shaft by either a socket or a flat tang, which was designed to break when the javelin hit its objective. As a result, when a clash was over, the Roman soldiers could recuperate the wooden shafts from the battlefield and re-use them. The pilum, in most cases, had a spike on the end of the shaft that made it easier to dig into the ground. Sometimes, in addition, the javelin could be weighted by a lead ball in order to increase its power of penetration. The maximum range of a pilum was approximately 35 metres, although the effective range was 15 to 20 metres.

The cavalry and light infantry of the auxilia were not equipped with the pilum, but with a lighter javelin known as *verutum*. The shaft of it was about 1.1 metres long. The tapering point of the verutum measured about 13cm and had inferior armour-piercing capabilities when compared with that of the heavier pilum. The verutum had a very simple design, since it lacked both the shank and the tang that could be found in the pilum.

Both the legionary cavalry and several mounted units of auxilia were equipped with a thrusting spear known as *hast*: this had an iron point with elongated shape – with the edges curving inwards from the belly of the blade to its tip – and a wooden shaft. The average length of such weapons was of 2 metres, including a butt spike made of iron.

A pugio with decorative scabbard of the type used by officers. (Legio XIII Gemina)

Swords

The *gladius* (short sword), the most iconic and important weapon of the Roman Army, was not Roman at all. The true origins of this arm are much clearer if we call it with its complete and proper name, the one used by the same Romans, 'gladius Hispaniensis'. The gladius was first designed in Iberia, in the territories of modern Spain and Portugal, as confirmed by various Roman sources and also by archaeological evidence. It was created by the Celtiberians, one of the many warrior peoples inhabiting Iberia during the Iron Age. Unlike other Iberians communities the Celtiberians were of mixed descent, being the product of Celtic migrations across the Iberian Peninsula. Because of their Celtic heritage, the Celtiberians produced weapons with very innovative techniques. The Romans abandoned their traditional swords of Greek fashion after the Second Punic War, as a result of their many encounters on the battlefield with the Celtiberian allies/mercenaries of Hannibal. They decided to adopt the same short swords as their Celtiberian opponents soon after realising that the gladius had great tactical potential.

According to the latest metallurgical studies conducted on surviving Roman short swords, the gladius could be forged following two different kinds of manufacture: it could be produced from a single piece of steel, or as a composite blade. Swords produced with the first process were created from a single bloom by forging from a temperature of 1,237°C. The carbon content increased from 0.05–0.08 per cent at the back side of the sword to 0.35–0.40 per cent on the blade, from which we can deduce that some form of carburisation may have been used. Swords produced with the second process were crafted by the pattern-welding process from five blooms reduced at a temperature of 1,163°C. Five strips of varying

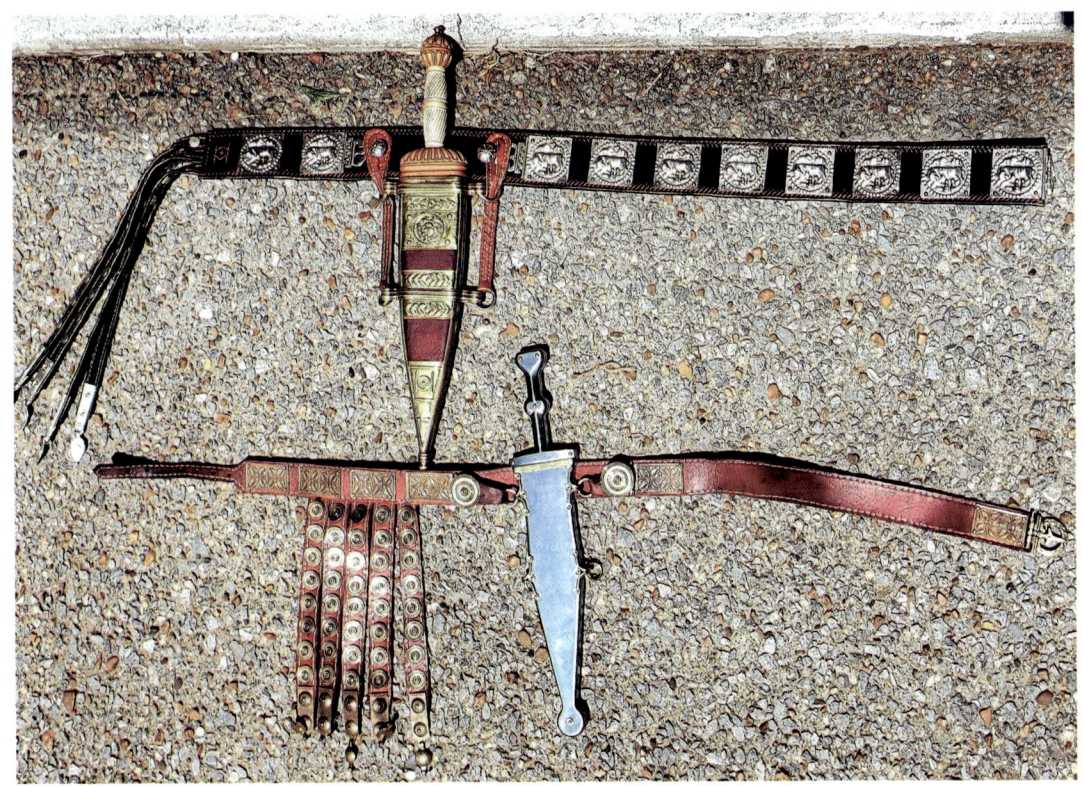

Two different models of pugio and *cingulum militiae*; the one with plain scabbard was used by legionaries, while decorated scabbard was used by officers. (Legio XIII Gemina)

carbon content were created: the central core of the sword contained the highest concentration of carbon (0.15–0.25 per cent); on its edges were placed four strips of low-carbon steel (with concentration of 0.05–0.07 per cent) and the whole thing was welded together by forging on the pattern of hammer blows. Each blow increased the temperature enough to create a friction weld at that spot. The forging operation, the most important one of the whole process, continued until the steel was cold, producing some central annealing. When produced by welding different strips together, the gladius had a channel down the centre of the blade; when produced by fashioning a single piece of steel, the blade had a rhomboidal cross-section. The blade of the gladius was two-edged for cutting and had a tapered point for stabbing during thrusting. A solid grip on the weapon was provided by a knobbed wooden *capulus* (hilt) added on the blade, usually having ridges for the fingers. The hilt could be decorated in many different ways; the swords of high officers and Praetorians, for example, usually had a hilt sculpted to resemble the head of an eagle.

The gladius continued to be produced over a very long time span, in several different designs. According to the traditional subdivision employed by military historians and archaeologists, the various types of gladius can be grouped into three main models: *Mainz*, *Fulham* and *Pompeii* (all named after the location where the canonical prototype of each group was actually found). The original Iberian sword, used from around 200BC until 20BC, had a slight 'wasp-waist' or 'leaf-blade' curvature that made it quite different from the following models. It was the largest and the heaviest model of gladius ever produced, with a blade length of 60–68cm and a sword length of 75–85cm. The blade was 5cm wide, with the overall weight of the weapon being 900g.

This earliest form of short sword, still heavily influenced by the original Iberian design, was used for a very long period when compared to its direct successors. The Roman city of Mainz was founded, like many other Roman cities, as a permanent military camp named Moguntiacum, around 13BC. The original military camp soon became an important centre for the production of swords and of other military equipment. With the transformation of the camp into a city, the manufacture of swords became even more significant, leading to the creation of a new kind of gladius, which is commonly known as the *Mainz gladius*.

The Mainz gladius kept the curvature of the previous model, but shortened and widened the blade. In addition, it modified the original point into a triangular one, specifically designed to thrust. The geographical diffusion of this model was limited to the border garrisons serving on the northern frontiers, and was different to the less battle-effective *Pompeii* version that came into use in other areas of the Empire. The short swords produced at Mainz during the early Imperial period were employed by the legions serving in the north, along the limes, but not exclusively so. Large numbers of these weapons were exported and sold extensively outside the boundaries of the Roman Empire. Apparently, various veteran ex-legionaries who had served on the frontiers used their discharge bonus on retirement to set up businesses as manufacturers and dealers of arms.

The *Mainz* variety of the gladius was characterised by a slight waist running the length of the blade. The average Mainz gladius had a blade length of 50–55cm and a sword length of 65–70cm. The blade was 7cm wide, with the overall weight of the weapon was 800g.

The *Fulham gladius* derives its name from the fact that the original prototype of this model was dredged from the River Thames near Fulham. The Fulham gladius had a slightly narrower blade than the Mainz variety, but the main peculiarity of this model consists of its triangular tip. It had a blade length of 50–55cm and a sword length of 65–70cm. The blade was 6cm wide, with the overall weight of the weapon being 700g.

The *Pompeii gladius* was the most popular of the three models that started to be produced after the original Iberian one. It had parallel cutting edges and a triangular tip. From a structural point of view, the Pompeii model eliminated the curvature, lengthened the blade and diminished the point. Regarding general dimensions, it was the shortest model of gladius: this was the result of the battle experiences of the Roman Civil Wars, during which Roman armies fought against each other in large pitched battles. In these cruel internal conflicts, hand-to-hand tactics were employed by both sides and so the traditional Roman

A legionary's pugio and cingulum militiae. (Marc Seriol (@marcmarkhus_photo), Legio II Traiana Fortis – Cohors I Barcinonum, Barcino Oriens)

military superiority lost most of its usual efficiency. Having to fight against enemies equipped exactly like themselves, with heavy cuirasses and shields, the Romans had to develop a lighter and shorter version of their sword, designed to thrust mainly with the point and in very strict spaces. The average Pompeii gladius had a blade length of 45–50cm and a sword length of 60–65cm. The blade was 5cm wide, with the overall weight of the weapon being 700g. With the end of the Roman civil wars, the Pompeii gladius gradually got longer. The gladius was carried in a scabbard mounted on a belt or, more frequently, on a shoulder stripe. It was worn on the left side of the soldier's body and thus each legionary had to reach across his body to draw it. Centurions, to differentiate themselves from their soldiers, wore the gladius on the right side of the body.

The Roman cavalrymen, be they legionary horsemen or auxiliaries, were not armed with the gladius of the foot legionaries/auxiliaries but with a longer *spatha* (cavalry sword). Originally designed by the Celts, it had a distinctive elongated leaf shape. It was double-edged and had a square kink or shallow 'V' point (its sides were drawn at an angle of 45 degrees to the axis of the blade). The tang of the spatha – ie, the internal part of the handle, made of metal but covered with organic material – swelled sharply to a point of greatest width just below its centre. The *ricasso* – ie, the unsharpened length of blade placed just above the handle of the sword – was very short and had a notch that varied a lot in depth.

Sword handles were made of wood or leather and generally had the form of an 'X', thus continuing the pattern of the famous Celtic 'antennae' swords. The handle was completed by a pommel, which was connected to the tang with a rivet-hole. The proper blade measured from about 60 to 90cm long and was made of iron or steel. In most cases, blades had a broad neck, with the greatest width being usually low down towards the point. Over time, points were increasingly rounded: this kind of shape confirms the fact that these cavalry weapons were used for slashing and not for thrusting.

Roman long swords were transported into iron scabbards, richly decorated with incisions and/or bosses; scabbards reproduced the general shape of the blade and were constructed from two plates: the front one, slightly wider than the back one, was folded over the latter along the sides. Each scabbard was reinforced by a decorated band around the top and by a sculpted chip at the bottom. Scabbards were generally suspended on the right hip from a belt made of leather; the sword was suspended from the waistbelt by means of a metal loop located at the back face of the scabbard.

Daggers

In addition to the gladius or the spatha, each Roman soldier also had a short dagger known as *pugio*. This sidearm was intended as an auxiliary stabbing weapon, specifically designed for hand-to-hand combat. Like the gladius short sword, it was designed by the Celtiberians and later adopted by the Romans. The pugio had a quite large leaf-shaped blade, which usually narrowed from the shoulders to run parallel to about half of the length before narrowing to a sharp point. A midrib ran close to the length of each side, either standing out from the face or sunken by grooves on each side. The tang was wide and flat; the grip was riveted through it, as well as through the shoulders of the blade. The pommel was originally round, but it soon assumed a new trapezoidal shape. Each pugio was transported in its own sheath made of soft metal, which had four suspension rings and a bulbous terminal expansion that was pierced by a large rivet. It developed from the original design over time: a rod tang was introduced and the hilt was no longer riveted through the tang since it started to be secured only at the shoulders of the blade. The hilt continued to consist of two layers of horn or wood. Each pugio was 18–28cm long and 5cm wide. The sheath could be heavily decorated, especially for the daggers used by officers or by Praetorians.

Military pack

Each legionary carried a *sarcina* (military pack), which comprised several fundamental pieces of equipment. Since the days of Gaius Marius's military reform, the Roman heavy infantrymen became famous for his ability as a combat engineer. In fact, infantrymen were capable of building fortified camps or military infrastructures such as bridges, in every condition and on every kind of terrain. The sarcina consisted of a *furca* (carrying pole), to which the following objects were suspended: a *loculus* (leather satchel), a waterskin, a small bag with rations of food for 15 days, one *patera* (shallow dish), one cooking pot, one skewer, some *sudis* (pointed stakes) used to construct fortifications, one *dolabra* (pickaxe), one *ligo* (mattock), one falx used to clear overgrowth, one *batillum* (shovel), and a basket for hauling earth. The equipment of a single legionary was completed with the *cingulum militare* (military belt), which had a very important symbolic significance: wearing it meant being part of the army, since civilians used different models of belt. Depriving a soldier of his military belt meant expelling him from the army.

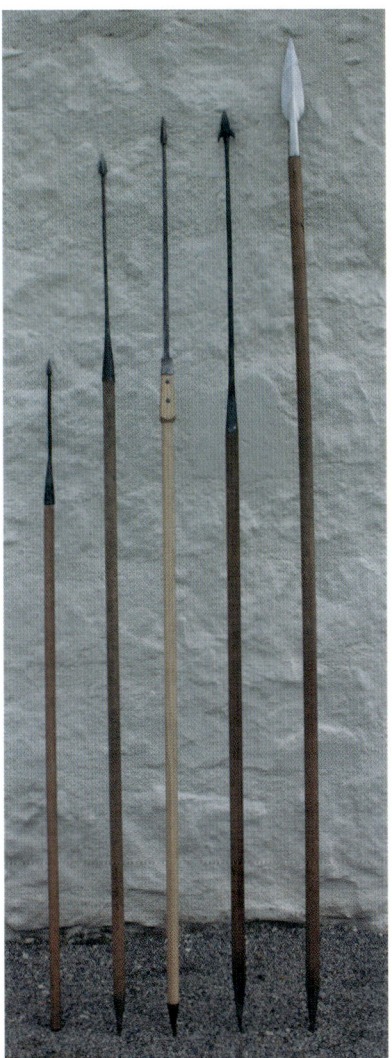 

Above left: Different throwing weapons employed by the legionaries. (Legio XIII Gemina)

Above right: Detail of the *pilum* and of the *sarcina*, or marching pack of the legionaries. (Legio XIII Gemina)

Basically the cingulum militare was a leather belt decorated with iron fittings and consisting of the following components: the *baltea* (hanging band), the *bulla* (rivets) that were applied on the baltea, the *pensilium* (pendants) that were applied at the end of the belt's straps, the *lamna* (disc) that was applied at the end of each apron strip embracing the pensilium and the *fibula* (buckle) of the belt.

Evolving equipment

During the course of the third century the traditional scutum used by the legionaries disappeared together with the lorica segmentata. The mass production of the scutum was complex, considering the various phases that were needed to create a shield of good quality. As for other components of the standard Roman military equipment, the production of the rectangular shield proved to be unsustainable for the productive structures and supply system of the Empire during the chaotic years of the third century AD. As a result, by the time of Constantine, almost all Roman military units had adopted a new and simpler model of oval shield. This was also carried by light infantry and cavalry, although with some possible variations in dimensions (the shields of light infantry and cavalry were generally a bit smaller than the heavy infantry ones, having a more circular shape).

The new oval shield was not convex but could be either dished (bowl-shaped) or flat. This difference in shape had some consequences for tactics, but it seems that the new model of shield was sometimes used to form the traditional defensive formations. The oval/round shield was much larger than the previous scutum and was constructed in a different way. In fact, it was made of solid planks instead of plywood and was supported by a double grip (at elbow and at hand, while the old scutum had just a single central grip). The general features of the oval/round shield made it much easier to produce than the previous rectangular one. In many ways, it was the direct heir of the oval shield used by the auxilia for its entire history. The new oval shield was about 110cm high and 90cm wide, being constructed with 1cm-thick wood planks; it was covered and bound with leather. A hollow iron or bronze boss covered the central hand grip. The former scutum was usually painted with standard decorations, which were almost identical for each legion. The shields of the auxiliaries were generally painted with standard motifs. It was not possible, however, to distinguish units from each other by the decorations on the shields.

The traditional heavy pilum was progressively abandoned around AD250. Unlike other pieces of the traditional Roman military equipment, it was not substituted by a single new weapon. The Roman infantrymen of the Middle Empire, in fact, could use two different kinds of javelin to break up the ranks of the enemy. Both these new missile weapons, known as *spiculum* (heavy javelin) and *verutum* (light javelin), had butt-spikes that protected the base of the shaft from rot and could be used for thrusting. The spiculum replaced the pilum as the heavy infantryman's main throwing weapon. It may have evolved from the gradual combination of the pilum with the standard Germanic javelin, known as *angon*. The latter was quite similar to (and probably derived from) the pilum, having a barbed head and a long narrow socket or shank, made of iron mounted on a wooden haft. The barbs were designed to lodge into the enemy shields so that the javelin could not be removed. The long iron shank prevented the head from being cut from the shaft, leaving combatants vulnerable and disrupting enemy formations.

The shaft may have been decorated and iron, or sometimes bronze, rings were fitted to it, which may have marked the centre of balance and the best place to hold the weapon. The Roman spiculum had all the main features of the Germanic angon: the main difference between the spiculum and the former pilum was the length of the thin point, because the new javelin tended to have a much shorter point than the previous one. In general, the spiculum was significantly shorter than its ancestor, perhaps being 190cm long. Its exact design is not fully known, because there were many variants. It had a medium iron shank attached to the head. In addition, the soldiers of the Middle Empire could also employ a lighter and longer-ranged javelin known as *verutum*. This was the main missile weapon of light infantry and cavalry, but could also be used by the heavy infantrymen in combination with the spiculum. The verutum had been used for skirmishing purposes since the days of the Republic, being issued to the light infantrymen of the legions. During the Crisis of the Third Century, however, it started to be employed by all kinds of units. The shaft of the verutum was about 1.1 meters long, thus being much shorter than that of the old pilum. Its point measured about 13cm; each verutum had either an iron shank like the spiculum or a tapering metal head. In general terms the verutum had a narrow armour-piercing head similar to that of the heavy javelins, though obviously with inferior capabilities of penetration. Over time, it seems that the spiculum started to be used on a lesser scale; apparently it was replaced by the verutum also as the main javelin for heavy infantry.

Since the reforms of Gaius Marius, the Roman heavy infantry had abandoned the spear as one of its main offensive weapons. All legionaries, in fact, started to be equipped with the pilum. The infantry of the auxiliary units, in contrast continued to employ the standard model of spear known as *lancea*. In the third century, a good number of the legionaries started to adopt the lancea of the

Roman legionaries training with wooden short swords. (Legio XI C.P.F. Hispaniensis)

Roman legionaries deployed for an inspection. (Marc Seriol (@marcmarkhus_photo), Legio II Traiana Fortis – Cohors I Barcinonum, Barcino Oriens)

auxiliaries as their main weapon. As a result, after several centuries, the Roman heavy infantry was again equipped with spears on a large scale. The lancea of the Middle Empire was different to that previously employed by the auxiliaries: it was light enough to be thrown just prior to contact but could be retained for use in hand-to-hand combat. Judging from the primary sources that are available, the heavy infantrymen of the Middle Empire preferred using the lancea for close fighting. Each soldier usually carried two or three spicula that were to be used as throwing weapons. If this was the situation for heavy infantry spears, we should bear in mind that the Roman cavalry had always continued to employ spears.

During the Crisis of the Third Century, however, the Roman heavy cavalry abandoned the traditional model of lancea and replaced it with a new one known as *contus*. This was a type of long wooden lance, originally used by the Iranian heavy cavalry. Since the first century AD, this heavy cavalry weapon started to be employed on a massive scale by the Sarmatians. It was later adopted also by the Parthian heavy cavalrymen, who employed it against the Romans.

The contus was about 4m long and had to be wielded with two hands while directing the horse using the knees; this made it a specialist weapon that required a lot of training and good horsemanship to use. Initially, only highly trained cavalrymen such as those fielded by the Sassanids could use such weapons in an effective way. The contus was reputedly a weapon of great power, especially when compared to other cavalry spears of its time. The great length of this lance was probably the origin of its name, since the Greek word *kontus* meant oar or barge-pole. The Roman cavalry introduced the contus on a large scale after facing the Sarmatian heavy cavalry in battle. The light cavalry, instead, continued to be armed with javelins of the verutum model.

Another piece of cavalry equipment copied by the Romans from the Sarmatians was the so-called *draco*. This was a military standard introduced into the Roman cavalry by Sarmatian auxiliaries during the second half of the second century AD. Apparently, it soon had a great success and was later adopted by infantry units. In the infantry, it substituted the traditional vexillum as the standard of the cohort. As a result, the Vexillifer of each cohort was replaced by a Draconarius. The eagle remained the primary symbol of the legions.

The draco was a traditional element of the Dacian military equipment, but it had been later adopted by the Sarmatians who passed its use to the Romans. As its name suggests, it had the form of a dragon with open wolf-like jaws containing several metal tongues. The hollow head of the dragon was metal and was mounted on a pole; it had a long fabric tube affixed at the rear. When used, the draco was held up into the wind where it filled with air and made a shrill sound as the wind passed through its metal tongues. As a result, in addition to being a military standard, it was a psychological weapon that resembled a terrifying dragon.

Clothing

In general terms, it is not correct to speak of uniforms for the armies of Antiquity, but we can surely say that the Roman Army of the Early and Middle Empires had a high level of standardisation in its dress. This was mainly due to the fact that state factories produced military clothing, equipment and weapons for all the soldiers of the Empire.

Tunic

The main garment worn by the Roman soldier was the tunic: this was short-sleeved but was produced in a version with long sleeves known as *tunica manicata,* which was of common use in the northern provinces of the Empire. The Roman tunic, made of wool for cold months and of linen for hot months, was a simple T-shaped garment reaching the knees of its wearer. It was not made of cut-and-sewn parts, but was woven entire in a single piece. Each tunica was folded in half and sewn up the sides, with a neck slit cut in the centre. The standard colours for military tunics were red and white; with white, in particular, was extremely popular in the hottest regions of the eastern provinces.

Cloak

After the tunic, the cloak was the second most important garment used by the Roman soldier. Made of wool and containing natural oil in order to repel water, it could be of two different kinds known as *sagum* and *paenula*. The first, typical of the Italic peninsula, was rectangular in shape; the second, instead, was rounded in shape and had a *cuculia* (hood) for head protection. All cloaks reached the hips and were double-folded so that they did not fall beneath the knees when worn. The cloaks of officers were quite different from those of soldiers, since they were longer (ankle-length) and trapezoidal in shape. All cloaks were pinned up on the right shoulder by a large *fibula* (brooch); this could be richly decorated if worn by a senior officer.

Trousers

The trousers, known as bracae, were considered by most of the well-educated Roman citizens as a barbarian garment; the legions garrisoning the northern frontiers, however, adopted them on a large scale from the early days of Empire.

Scarf

All legionaries wore a *focale* (scarf) around their neck; this was made of wool for cold months and of linen for hot months. The focale protected the neck from chafing by the armour and was loosely knotted on the front.

Footwear

For footwear, all Roman soldiers wore the famous open sandals known as *caligae* (derived from Emperor Caligula, who loved them). These were made from leather and were laced up the centre of the foot as well as on to the top of the ankle. Iron hobnails were hammered into their sole for added strength.

Dress denoting rank

Centurion

A centurion was distinguished by carrying the gladius on the left side and the pugio on the right side (the legionaries did the opposite). He had a peculiar transverse crest on the helmet. He always carried a *vitis* (vine-stick) that was used to mete out punishments. He applied some sculpted discs known as *phalerae* (made of gold, silver or bronze) on the front of his armour. The phalerae were military medals, usually awarded for distinguished conduct in action. They were aligned on three horizontal lines and were attached to orthogonal leather straps in order to form a single breastplate that was worn on the chest.

Optio

An optio was easy to recognise since he had plumes of horse hair or feathers applied on each side of his helmet, with the standard crest in the middle. He carried a *hastile* (long wooden staff) that was as tall as the optio himself and had a round metal pommel at its end. Like the vine-stick of centurions, this was used to mete out punishments.

Standard-bearers and musicians

All standard bearers carried a peculiar small round shield and wore a fur over their shoulders/head: the latter was a wolf one for the 'vexillifer', a bear one for the 'signifer' and a lion or tiger one for the

Different working tools employed by the legionaries. (Legio XIII Gemina)

'aquilifer'. The fur was worn as a cape, with the head of the animal strapped to the helmet of the standard bearer. In many cases, the standard bearers, being veterans, wore on their chests the same phalerae of the centurions. The military musicians of a legion were usually equipped like the standard bearers and carried small round shields and wore a fur over their shoulders/head.

Evolving dress

The general appearance of the Roman soldiers during the Crisis of the Third Century became progressively quite different from that reproduced on the famous Trajan's Column. The traditional image of the legionaries equipped with segmented armour and of the auxiliaries equipped with chainmail was no longer in existence by AD284. The Crisis of the Third Century saw a progressive collapse of the Roman central administration. This element had important consequences for the organisation of the military forces and for the supply system that had worked very well until that time. During those years of anarchy, the regular soldiers started to receive new clothes and weapons on a much more irregular basis; this resulted in the partial loss of the high level of standardisation that had been reached previously.

During the Middle Empire, the main garment worn by the Roman soldiers was the tunic. However, this had changed a lot since the times of Augustus. The Arch of Severus, which was completed at the beginning of the third century, shows Roman soldiers with the traditional tunic worn during the Early Empire. By the central decades of the third century, the new long-sleeved tunic was widespread. This was copied from the contemporary dress of the Germanic warriors, who already used it on a large scale.

Tunics were generally quite decorative, with woven or attached strips known as *clavi* and circular roundels of cloth called *orbiculi*. Both these decorative elements usually comprised geometric patterns and stylised plant motifs but could include human or animal figures. The clavi were narrow vertical bands of cloth running from the top to the bottom of the tunic, while the orbiculi were patterned roundels located on the shoulders and skirt of the tunic.

Left: A side view of the legionary *caliga*, or military sandal. (Marc Seriol (@marcmarkhus_photo), Legio II Traiana Fortis – Cohors I Barcinonum, Barcino Oriens)

Below: The Roman military camp. (Legio XIII Gemina)

After the tunic, the cloak remained the second most important garment of the Roman soldier; hip-length and semi-circular, it was known as *sagum*. It could be decorated with fringes or with applied decorations of the same kind worn on the tunic. The cloaks of officers were quite different to those of soldiers. They were longer (ankle length) and trapezoid in shape and featured large embroidered panels in the form of a square known as *tabula*. All cloaks were pinned up on the right shoulder by a large brooch, which could be richly decorated, if worn by a senior officer. Materials and colours were more or less the same as for the tunics: wool for winter and linen for summer, red and white being the most popular colours.

During the late third century, Roman military belts started to show a strong Germanic influence, especially in the decorations: bronze or iron fittings became increasingly popular, both cast and chiselled, being embellished with punched or carved work. By the time of Diocletian and Constantine, most of the soldiers wore bracae.

Regarding footwear, the traditional caligae had completely disappeared by the end of the third century. They had been replaced by new front-fastening short boots with integrally cut laces.

When not wearing the helmet, soldiers on active service usually had a particular kind of cap on their heads known as *pileus pannonicus*. This could be made from felt or napped wool; it could be smooth or shaggy, low or tall. The cylindrical shape made this kind of cap very comfortable and easy to produce. It started to be used during the early third century, being adopted as a sign of distinction also by civilian dignitaries. This pillbox cap was usually covered with fur in the northern regions of the Empire.

Battle tactics

During the Early Empire, Roman battle tactics were highly standardised and incredibly effective. The first weapon to be used in a pitched clash was the pilum. Once the enemy ranks had been shattered by the initial rain of javelins, the legionaries drew their short swords and charged their opponents. According to the Roman tactical doctrine, emphasis was put on using the scutum to provide maximum body coverage, while the gladius was used to attack with devastating thrusts and short cuts. With these tactics, the Romans were able to defeat any enemy infantry for centuries. In addition, this kind of warfare limited the number of casualties suffered by the imperial troops. Using their swords to thrust in the few spaces created between the shields of their close formations, the legionaries were rarely exposed to the offensive weapons of the enemies, who had very few chances to manoeuvre.

Intensive and continuous training made the Roman legionaries efficient like machines: the stabbing wounds produced by their short swords were almost always mortal, especially if hitting in the abdominal area (intended as the main target for thrusts since the early phase of a legionary's intensive training). Each Roman infantryman was trained to adapt to any possible combat situation: each one of his weapons could be used in different ways and he had to be ready to exploit fully any mistake of the enemy or any favourable momentum. For example, Roman legionaries were trained to slash kneecaps beneath their shield wall or to cut the throat of the enemies while charging in the famous *testudo* formation. In the latter, the legionaries aligned their rectangular shields to form a packed rectangle covered on the front and on the top. The first row of men held their shields from about the height of their shins to their eyes, in order to present a 'shield-wall' to all sides. The men in the back ranks placed their shields over their heads to protect the formation from above and balanced the shields on their helmets by overlapping them. The legionaries on the sides and rear of the testudo held their shields like those of the first row. The 'turtle' formation of the Roman legions offered excellent protection against the missile weapons of any possible opponent.

Roman legionaries on the march. (Legio XIII Gemina)

During the glorious days of the Early Empire, the Romans usually employed a general forward defence strategy in which the legions, stationed on the frontiers of the Empire, were prepared to neutralise imminent enemy incursions or invasions before they could reach the territories of Rome. This kind of strategy had proved to be very efficient on several occasions, mainly thanks to the creation of strategic salients beyond the limes. This system of defence, however, had been created to counter the low-intensity raids of the Germanic tribes or the invasions of regular Parthian/Sassanid armies: it was not suited to stop mass migrations of entire peoples, like those that occurred on the Rhine and Danube frontiers starting from the reign of Marcus Aurelius.

At the end of the second century AD, the Roman military machine was completely surprised by this new kind of threat: the generals of Marcus Aurelius had to struggle for years against the Germanic tribes in order to repulse their invasions, suffering very heavy human losses that shattered the same basic structure of the legions. After these dramatic events, it soon became clear to the imperial military leaders that a diffused defence of the borders was no longer the right response to foreign menaces. As a result, Gallienus and all following monarchs started to create large reserves of highly mobile troops that were stationed around the major cities of the Empire.

The 'forward defence' was by now too vulnerable: the migrating Germanic tribes could easily concentrate thousands and thousands of warriors in a precise point of the limes, which was usually defended by a single legion with its relative auxiliary units. The lack of reserves to the rear of the border meant that a foreign invading force could penetrate very deeply into the Empire before Roman reinforcements from other border garrisons could arrive to intercept it. The new defensive strategy created by Gallienus and developed by his successors is commonly known as 'defence in depth'. The Romans accepted that their

Roman legionaries throwing their javelins. (David Burns, Legio XIII Gemina)

frontier provinces would become the main combat zone in the military operations conducted against the barbarians, rather than the enemy lands located across the border. If the previous 'forward defence' had been based on the principle that the best defence is preventive offence, the new strategy was much more passive and definitively abandoned any possibility of territorial expansion for Rome.

Evolution of tactics

The most important change of the third century was the great ascendancy of cavalry: for centuries, since the days of the Greek hoplites who fought against the Persians in the fifth century BC, the armies of the Mediterranean world had seen a clear prominence of the infantry over the cavalry. The same Romans had conquered immense territories using the power of their legions, exporting their political control thanks to the superiority of the legionary heavy infantryman. But the epochal events happening during the Middle Empire were to change all this, leading to the progressive decline of infantry warfare. The contacts with new military civilisations, like those of the Sarmatians and Sassanids, led the Romans to the adoption of new tactical principles. These were mostly employed, as a response to the new military threats faced since the third century, so they were strongly linked with the adoption of the new 'defence in depth' strategy.

As we have already seen, under Gallienus cavalry increased significantly from a numerical point of view: until that time, in fact, it had been just a secondary element of the Roman military apparatus,

Roman legionaries advancing in *testudo*, or turtle, formation. (David Burns, Legio XIII Gemina)

being formed only by small contingents of legionary cavalry or by units of auxiliaries. Taking over tactical prominence from the infantry did not just have numerical consequences: the general status of the mounted forces was deeply affected by the tactical changes. Roman generals of the Middle Empire usually preferred to avoid large pitched battles: unlike those of previous centuries. In this era, the Roman forces usually tried to avoid direct confrontations with large enemy armies, if possible. This was mainly due to recruiting problems and to the economic difficulties of the Empire. In order to minimise casualties, the Roman generals had several tactical options at their disposal: night attacks, ambushes, surprise raids, harassment and strategic manoeuvring.

Bibliography

Primary sources

Appian, *Roman History*

Appianus, *Gallic History*

Appianus, *Illyrian Wars*

Appianus, *Wars in Spain*

Dio Cassiuss, *Roman History*

Diodorus Siculus, *History*

Diodorus Siculus, *Library of History*

Dionysios of Halikarnassos, *Roman Antiquities*

Livy, *History of Rome from its foundation*

Plutarch, *Lives*

Polybius, *The Histories*

Strabo, *Geography*

Tacitus, *The Histories*

Tacitus, *Agricola*

Tacitus, *The Annals*

Secondary sources

Allen S, *Celtic Warrior 300BC–AD100*, Osprey Publishing, 2001

Anderson E B, *Cataphracts: Knights of the Ancient Eastern Empires*, Pen & Sword, 2016

Baker P, *Armies and Enemies of Imperial Rome*, Wargames Research Group, 1981

Bishop M C, Coulston J C, *Roman Military Equipment from the Punic Wars to the Fall of Rome*, Oxbow Books, 2006

Brzezinski R and Mielczarek M, *The Sarmatians 600BC–AD450*, Osprey Publishing, 2002

Connolly P, *Greece and Rome at War*, Macdonald Phoebus Ltd, 1981

Cowan R, McBride A, *Imperial Roman Legionary AD161–284*, Osprey Publishing, 2003

Cowan R, O'Brogain S, *Roman Guardsman 62BC–AD324*, Osprey Publishing, 2014

Cowan R, O'Brogain S, *Roman Legionary AD284–337*, Osprey Publishing, 2014

Elliot P, Legions in Crisis: *Transformation of the Roman Soldier AD192–284*, Fonthill Media, 2014

Farrokh K, *The Armies of Ancient Persia: the Sassanians*, Pen & Sword, 2014

Goldsworthy A, *The Fall of the West: The Slow Death of the Roman Superpower*, Weidenfeld & Nicolson, 2009

Gorelik K, *Warriors of Eurasia*, Montvert Publishing, 1995

Kiley K F, *An Illustrated Encyclopedia of the Uniforms of the Roman World*, Lorenz Books, 2012

Macdowall S, Embleton G, *Late Roman infantryman AD236–565*, Osprey Publishing, 1994

Macdowall S. Hook C, *Late Roman cavalryman AD236–565*, Osprey Publishing, 1995

Macdowall S, McBride A, *Germanic Warrior AD236–568*, Osprey Publishing, 1996

Newark T, *Ancient Armies*, Concord Publications, 2000

Newark T, *Ancient Celts*, Concord Publications, 1997

Newark T, *Barbarians*, Concord Publications, 1998

Newark T, *Warlord Armies*, Concord Publications, 2004

Nicolle D, McBride A, *Rome's Enemies 5: the Desert Frontier*, Osprey Publishing, 1991

Quesada Sanz F, *Armas de Grecia y Roma*, La Esfera, 2014

Rankow B, Hook R, *The Praetorian Guard*, Osprey Publishing, 1994

Rocca S, *The Army of Herod the Great*, Osprey Publishing, 2009

Rostaing N, *Lost Pontic legions: Pontic imitation legions used by Mithridates*, Ancient Warfare Magazine, volume X, issue 3

Simkins M, Embleton R, *The Roman Army from Hadrian to Constantine*, Osprey Publishing, 1979

Sumner G, *Roman Military Clothing (1): 100BC–AD200*, Osprey Publishing, 2002

Sumner G, *Roman Military Clothing (2): AD200–400*, Osprey Publishing, 2003

Travis J, Travis H, *Roman Body Armour*, Amberley, 2012

Travis J, Travis H, *Roman Helmets*, Amberley, 2015

Travis J, Travis H, *Roman Shields*, Amberley, 2014

Warry J, *Warfare in the Classical World*, Salamander Books, 1997

Webber C, *The Thracians 700BC–AD46*, Osprey Publishing, 2001

Wilcox P, *Rome's Enemies 1: Germanics and Dacians*, Osprey Publishing, 1982

Wilcox P, *Rome's Enemies 2: Gallic and British Celts*, Osprey Publishing, 1985

The Re-enactors Who Contributed to This Book

Ala I Batavorum

There are clouds of dust on the horizon, slowly heading in your direction. It turns out to be a squad of Roman cavalrymen, fully armed, armour shining and a highly decorated horse. The approaching horsemen wear silver mask, blinking in the sun, their faces covered, only showing their eyes: spectacular, yet terrifying. Stichting Ala I Batavorum recreates these cavalrymen from ages gone by. In our regular show, our fully equipped cavalrymen will show their skills at arms on horseback. Equipment and tactics will be shown and explained, both in single file as well as operating in the group. All equipment and the tactical advantage of the cavalry within the Roman army will be explained to the public. This entertainment is a spectacular and educational show. Stichting Ala I Batavorum can bring together up to six cavalrymen. All our equipment is based upon the archaeological records and is well researched. Our impressions focus mainly on the Batavian auxilia cavalry, but also include Augustean and late Roman impressions. Most impressive is our late Roman *clibinarius* kit, incorporating a recreation of the Dura Europos (Syria) horse scale armour.

Contacts
E: info@ala-batavorum.nl
W: www.ala-batavorum.nl
Facebook: https://m.facebook.com/ALA-BATAVORUM-133726753365976/

Legio II Traiana Fortis – Cohors I Barcinonum, Barcino Oriens

Barcino Oriens is a Roman historical re-enactment, reconstruction and disclosure group established in 2009, at the Joan Pelegrí School in Barcelona. Its purpose is to spread the knowledge of ancient Roman history so that everyone can see, feel and touch it. The group presents activities that bring us closer to the reality of the ancient Roman world. Each historical reconstruction is presented in the closest way to the interlocutor so that the audience develops enough historical empathy. Additionally, the objects, images, tools, costumes and various realities of antiquity are brought closer to the viewer. The purpose of the recreation and historical reconstruction of Barcino Oriens is to link and let people know, in a direct, attractive and emotional way, our reality to that of our origins, to that of the Roman Barcino. With Legio II Traiana Fortis – Cohors I Barcinonum we recreate and reconstruct differents aspects of a Roman legion. You can find out how the legionaries were recruited and trained, how they lived in a Contubernium, what the battle orders and instructions were, and much more. All historical reconstructions and re-enactments are accompanied by a rigorous historical and academic explanation. We are educators of Roman history spreading Catalonia's archaeological heritage. All our performances are theatrical stagings of different aspects of life in ancient Rome, from the daily life of the patrician ladies, through different aspects of military. It is a way of creating 'custom paintings' where these fragments of life are dramatised. All this is achieved through the performance of the members of the group, who create characters that are part of each painting. Each performance becomes a small play (called 'pill') lasting 30 to 45 minutes. We want the crowd to understand it and participate, immersing them in the atmosphere and the words of the characters, contemplating the material and costumes of our historical reconstruction. Ricard Llop Altés is the director of the group.

Contacts
E: barcinooriens@barcinooriens.cat
W: www.barcinooriens.cat
Facebook: https://www.facebook.com/barcinooriens/

Legio XI C.P.F. Hispaniensis
Legio XI CPF Hispaniensis was born on 7 July 2018, founded by veteran Spanish and Swiss re-enactors who wanted create a correct historical reenactment group in Spain. Currently, the group has 25 members and recreates COH III of Legio XI during the first century AD (Flavian times). The group is twinned with the Legio XI groups in Switzerland, Belgium and Russia. It is considered the best dressed group in Europe and has participated in events such as 'Augusta Raurica' or 'Natale di Roma' in addition to training and military ceremonies, marches and camps. Several of its members are Roman horsemen. All the group's activities are characterised by a high level of accuracy and professionalism.

Contacts
E: cohorslemav@gmail.com
Facebook: https://www.facebook.com/LEGXICOHIII/

Legio VI Victrix Cohors II Cimbria
Legio VI Victrix Cohors II Cimbria is a Scandinavian Roman Re-enactment group started in Denmark, and with a chapter in Sweden, and members from Germany. Our group consists of military and civilian reenactors. We re-enact the life in the Roman legions in the late first century AD and convey the life of the Roman Empire and its relations to the regions north of Germania Inferior, the Cimbrian peninsula and Scandia. Every summer, Cohors II Cimbria participates in events in Scandinavia and around Europe. In the 'off-season', we research, lecture and are develop our knowledge and equipment. Roman re-enactment is a never-ending process, as new archeological finds and new theories arise and constantly change our perspective. Scandinavia was never a part of the Roman Empire, but due to our many bog finds, most of which are from the Roman period 0–AD350, we have a lot of Roman artifacts including the largest collection of Roman imperial weapons outside of Italy. The tribes that lived here at that time – mainly the Cimbrians, Teutons, Charudes and Sviones – were, from the time of Augustus onwards, apparently on friendly terms with the Romans, or so the story goes. As you can read in the *Res Gestae of Augustus*, that primary source mentions an expedition to these most northern regions in year AD5 led by Tiberius. That is one of many reasons Roman re-enactment in Scandinavia is important. To us there is a forgotten story to be told.

Contacts
E: cohorsllcimbriadk@gmail.com
W: https://cohorsllcimbria.com/
Facebook: https://www.facebook.com/Cohorsllcimbria/

Legio XIII Gemina
Legio XIII Gemina or Exercitus Pannonia Superior – Legio XIII Gemina is an association dating back to 1992 and based in Vienna, Austri). As the name suggests, the main focus is on the display of military life

on the Danube border of legionary and auxiliary soldiers. We also try to recreate civilian life. Our chosen time period is the turbulent years at the end of the reign of Emperor Nero up until Emperor Domitian. During the latter's reign, Legio XIII built the legionary fortress of Vindobona (modern day Vienna). Our second military unit is Cohors II Italica Civium Romanorum, a rather interesting corps reconstructed according to a tombstone found at Carnuntum belonging to Proculus, an optio in this cohort and also a member of a Vexillatio of Sagittarii (archers) of the Syrian Army. Most probably this unit was sent to Pannonia during the civil wars of AD69, better known as the Year of the Four Emperors. We produce as much as of our equipment as possible, staying in constant contact with museums and archaelogists for the latest scientific research, and always trying to improve our artifacts and performances at public events. Our goal is to present a common Roman past, which stretches over wide parts of Europe and around the Mediterranean, to an interested audience at events in Austria, Germany, Switzerland, Italy, France and the Czech Republic.

Contacts
E: obmann@legxiii.at
W: http://www.legxiii.at
Facebook: https://www.facebook.com/Legio-XIII-Gemina-125111774228689

Legio XIIII Gemina

Legio XIIII Gemina Martia Victrix is a re-enactment and living history group located in the United States, which is dedicated to representing the Roman legionary as he may have appeared during the years of the emperors from Nero to Trajan. We are a Roman re-enacting legion, with members from the United States and Canada. Our group is based out of Ohio and surrounding states. We strive for authenticity with our impressions to better educate the general public. Founded in 2015, our group seeks to recreate the equipment, routines and the experiences of Roman legionaries from the first century AD for public, private and educational events.

Contacts
E: info@legxiiii.com
We: http://legxiiii.com/
Facebook: https://www.facebook.com/legioxiiii/?ref=page_internal

Other books you might like:

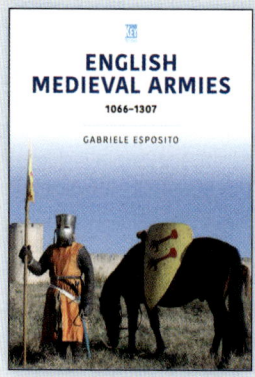

Historic Armies Series,
Vol. 1

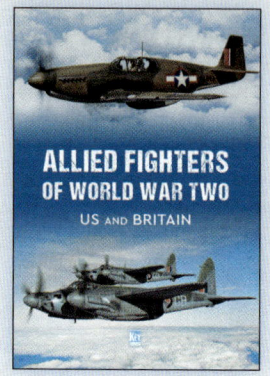

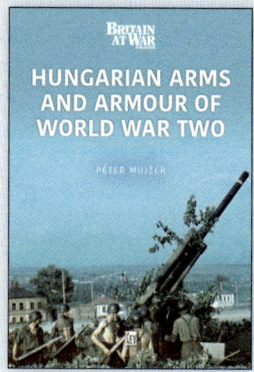

Military Vehicles and
Artillery Series,
Vol. 2

Military Vehicles and
Artillery Series,
Vol. 5

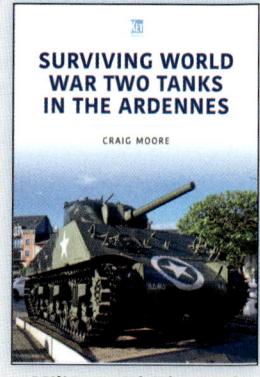

Military Vehicles and
Artillery Series,
Vol. 4

For our full range of titles please visit:
shop.keypublishing.com/books

VIP Book Club
Sign up today and receive
TWO FREE E-BOOKS

Be the first to find out about our forthcoming
book releases and receive exclusive offers.

Register now at keypublishing.com/vip-book-club

Our VIP Book Club is a 100% spam-free zone, and we will never share your email with anyone else.
You can read our full privacy policy at: privacy.keypublishing.com